Modeling Digital Switching Circuits with Linear Algebra

Synthesis Lectures on Digital Circuits and Systems

Editor
Mitchell A. Thornton, *Southern Methodist University*

The Synthesis Lectures on Digital Circuits and Systems series is comprised of 50- to 100-page books targeted for audience members with a wide-ranging background. The Lectures include topics that are of interest to students, professionals, and researchers in the area of design and analysis of digital circuits and systems. Each Lecture is self-contained and focuses on the background information required to understand the subject matter and practical case studies that illustrate applications. The format of a Lecture is structured such that each will be devoted to a specific topic in digital circuits and systems rather than a larger overview of several topics such as that found in a comprehensive handbook. The Lectures cover both well-established areas as well as newly developed or emerging material in digital circuits and systems design and analysis.

Modeling Digital Switching Circuits with Linear Algebra
Mitchell A. Thornton
2014

Arduino Microcontroller Processing for Everyone! Third Edition
Steven F. Barrett
2013

Boolean Differential Equations
Bernd Steinbach and Christian Posthoff
2013

Bad to the Bone: Crafting Electronic Systems with BeagleBone and BeagleBone Black
Steven F. Barrett and Jason Kridner
2013

Introduction to Noise-Resilient Computing
S.N. Yanushkevich, S. Kasai, G. Tangim, A.H. Tran, T. Mohamed, and V.P. Smerko
2013

Atmel AVR Microcontroller Primer: Programming and Interfacing, Second Edition
Steven F. Barrett and Daniel J. Pack
2012

Representation of Multiple-Valued Logic Functions
Radomir S. Stankovic, Jaakko T. Astola, and Claudio Moraga
2012

Arduino Microcontroller: Processing for Everyone! Second Edition
Steven F. Barrett
2012

Advanced Circuit Simulation Using Multisim Workbench
David Báez-López, Félix E. Guerrero-Castro, and Ofelia Delfina Cervantes-Villagẫmez
2012

Circuit Analysis with Multisim
David Báez-López and Félix E. Guerrero-Castro
2011

Microcontroller Programming and Interfacing Texas Instruments MSP430, Part I
Steven F. Barrett and Daniel J. Pack
2011

Microcontroller Programming and Interfacing Texas Instruments MSP430, Part II
Steven F. Barrett and Daniel J. Pack
2011

Pragmatic Electrical Engineering: Systems and Instruments
William Eccles
2011

Pragmatic Electrical Engineering: Fundamentals
William Eccles
2011

Introduction to Embedded Systems: Using ANSI C and the Arduino Development
Environment
David J. Russell
2010

Arduino Microcontroller: Processing for Everyone! Part II
Steven F. Barrett
2010

Arduino Microcontroller Processing for Everyone! Part I
Steven F. Barrett
2010

Digital System Verification: A Combined Formal Methods and Simulation Framework
Lun Li and Mitchell A. Thornton
2010

Progress in Applications of Boolean Functions
Tsutomu Sasao and Jon T. Butler
2009

Embedded Systems Design with the Atmel AVR Microcontroller: Part II
Steven F. Barrett
2009

Embedded Systems Design with the Atmel AVR Microcontroller: Part I
Steven F. Barrett
2009

Embedded Systems Interfacing for Engineers using the Freescale HCS08 Microcontroller
II: Digital and Analog Hardware Interfacing
Douglas H. Summerville
2009

Designing Asynchronous Circuits using NULL Convention Logic (NCL)
Scott C. Smith and JiaDi
2009

Embedded Systems Interfacing for Engineers using the Freescale HCS08 Microcontroller
I: Assembly Language Programming
Douglas H.Summerville
2009

Developing Embedded Software using DaVinci & OMAP Technology
B.I. (Raj) Pawate
2009

Mismatch and Noise in Modern IC Processes
Andrew Marshall
2009

Asynchronous Sequential Machine Design and Analysis: A Comprehensive Development
of the Design and Analysis of Clock-Independent State Machines and Systems
Richard F. Tinder
2009

An Introduction to Logic Circuit Testing
Parag K. Lala
2008

Pragmatic Power
William J. Eccles
2008

Multiple Valued Logic: Concepts and Representations
D. Michael Miller and Mitchell A. Thornton
2007

Finite State Machine Datapath Design, Optimization, and Implementation
Justin Davis and Robert Reese
2007

Atmel AVR Microcontroller Primer: Programming and Interfacing
Steven F. Barrett and Daniel J. Pack
2007

Pragmatic Logic
William J. Eccles
2007

PSpice for Filters and Transmission Lines
Paul Tobin
2007

PSpice for Digital Signal Processing
Paul Tobin
2007

PSpice for Analog Communications Engineering
Paul Tobin
2007

PSpice for Digital Communications Engineering
Paul Tobin
2007

PSpice for Circuit Theory and Electronic Devices
Paul Tobin
2007

Pragmatic Circuits: DC and Time Domain
William J. Eccles
2006

Pragmatic Circuits: Frequency Domain
William J. Eccles
2006

Pragmatic Circuits: Signals and Filters
William J. Eccles
2006

High-Speed Digital System Design
Justin Davis
2006

Introduction to Logic Synthesis using Verilog HDL
Robert B.Reese and Mitchell A.Thornton
2006

Microcontrollers Fundamentals for Engineers and Scientists
Steven F. Barrett and Daniel J. Pack
2006

Modeling Digital Switching Circuits with Linear Algebra

Mitchell A. Thornton

ISBN: 978-3-031-79866-5   paperback
ISBN: 978-3-031-79867-2   ebook

DOI 10.1007/978-3-031-79867-2

A Publication in the Springer series
SYNTHESIS LECTURES ON DIGITAL CIRCUITS AND SYSTEMS

Lecture #44
Series Editor: Mitchell A. Thornton, *Southern Methodist University*
Series ISSN
Print 1932-3166   Electronic 1932-3174

Modeling Digital Switching Circuits with Linear Algebra

Mitchell A. Thornton
Southern Methodist University

SYNTHESIS LECTURES ON DIGITAL CIRCUITS AND SYSTEMS #44

ABSTRACT

Modeling Digital Switching Circuits with Linear Algebra describes an approach for modeling digital information and circuitry that is an alternative to Boolean algebra. While the Boolean algebraic model has been wildly successful and is responsible for many advances in modern information technology, the approach described in this book offers new insight and different ways of solving problems. Modeling the bit as a vector instead of a scalar value in the set $\{0, 1\}$ allows digital circuits to be characterized with transfer functions in the form of a linear transformation matrix. The use of transfer functions is ubiquitous in many areas of engineering and their rich background in linear systems theory and signal processing is easily applied to digital switching circuits with this model. The common tasks of circuit simulation and justification are specific examples of the application of the linear algebraic model and are described in detail. The advantages offered by the new model as compared to traditional methods are emphasized throughout the book. Furthermore, the new approach is easily generalized to other types of information processing circuits such as those based upon multiple-valued or quantum logic; thus providing a unifying mathematical framework common to each of these areas.

Modeling Digital Switching Circuits with Linear Algebra provides a blend of theoretical concepts and practical issues involved in implementing the method for circuit design tasks. Data structures are described and are shown to not require any more resources for representing the underlying matrices and vectors than those currently used in modern electronic design automation (EDA) tools based on the Boolean model. Algorithms are described that perform simulation, justification, and other common EDA tasks in an efficient manner that are competitive with conventional design tools. The linear algebraic model can be used to implement common EDA tasks directly upon a structural netlist thus avoiding the intermediate step of transforming a circuit description into a representation of a set of switching functions as is commonly the case when conventional Boolean techniques are used. Implementation results are provided that empirically demonstrate the practicality of the linear algebraic model.

KEYWORDS

digital logic, digital design, transfer function, digital logic model, switching circuit, switching theory, linear algebra, justification, implication, electronic design automation, spectral methods, BDD, binary decision diagram, MVL, multiple-valued logic, SAT, satisfiability, ATPG, automatic test pattern generation, Reed-Muller, Walsh, Chrestenson

Contents

CHAPTER 1

Introduction

Boolean problems are conventionally formulated and solved using the algebraic framework originally proposed by George Boole for application to symbolic logic problems. Boole's algebra was used by Claude Shannon to model networks of switching relays and later, for the more general manipulation of information in the form of binary digits, or bits. An alternative approach is to model data as an element within a vector space. This approach is motivated by the notion of a quantum bit, or qubit, as used in the quantum computing community.

The vector space model allows the framework of linear algebra to be used for solving problems instead of Boolean algebraic axioms and postulates. The vector space model for information results in switching circuit models as transformations among vector spaces in the form of a characterizing transformation matrix. Common *Electronic Design Automation* (EDA) problems solved using Boolean switching theory such as simulation and justification become solutions to linear algebra problems.

The vector space model provides a unifying mathematical framework for both binary and general multi-valued switching networks since each of these may be formulated using vectors to represent fundamental units of information. Boolean algebras are not functionally complete over sets of truth values with cardinalities other than a power of two, thus multi-valued switching network modeling has traditionally used alternative algebraic formulations. The vector space model overcomes this problem. Quantum logic is expressed using qubits that are mathematically modeled as vectors with complex-valued components and the method described here allows for the relationship between quantum and conventional electronic switching circuits to be easily observed.

The use of Boolean algebra is ever present in modern information processing tasks including both hardware and software design and implementation. Boolean algebra was proposed by George Boole for the purpose of manipulating symbolic logic expressions in his seminal work [1]. The application of Boolean algebra to modern information processing is credited to Claude Shannon who proposed its use in modeling networks of electrical relays in his 1937 Masters degree thesis at the Massachusetts Institute of Technology [2]. Shannon later generalized these ideas for the general modeling of information in communications networks during his tenure at Bell Laboratories [3] firmly implanting the notion of the binary digit, or bit, as the fundamental atomic unit of information. The use of the bit and its manipulation with the postulates and axioms of Boolean algebra has since become the common and pervasive methodology for modern information processing tasks and is heavily utilized in many areas such as data communications, software development, and digital circuit design. Switching theory is based upon the ideas of Shannon and comprises a vast amount of theory and techniques regarding the modeling and manipulation of modern transistor-based switching circuits for the purpose of information processing. There have been many advances in switching theory resulting in a large number of results and methods commonly used in modern digital integrated circuit design. Many volumes have been written about switching theory such as [31].

While traditional switching theory is certainly in part responsible for the information age of present, there is not a fundamental reason that information models must be restricted to the domain of Boolean algebra. Indeed, the emerging field of quantum computing utilizes the concept of a quantum bit, or qubit, that is modeled as an element or vector in a finite dimensioned Hilbert vector space. The vector model for the qubit arose from the work of John von Neumann who used a Hilbert space model for describing the laws of quantum mechanics [4]. The qubit model provides the motivation for investigating the use of a vector space model for conventional electronic digital circuitry and is the subject of this book.

In keeping with the notation commonly used in the emerging field of quantum logic and computing, we use the notation of Paul Dirac [5] and this will be briefly defined and correlated with the more common notation found in the general literature on linear algebra. The slight departure from quantum logic notation is that we shall represent the atomic data values as row vectors rather than column vectors as commonly used in the quantum computing community. While this choice is arbitrary in terms of the underlying mathematics, it has the advantage of more clearly illustrating the isomorphic relation between truth tables for Boolean functions and transfer matrices in the vector space domain. The key mathematical difference between quantum logic network models and conventional switching network models is that the transfer matrices describing them are not unitary. Rather the transformation matrices are in general not of full rank and often they are non-square. For this reason, generalized matrix inverses are required and we make use of the Moore-Penrose pseudo-inverse for our purposes [6].

The definition of the transformation matrix, or transfer matrix, describing a conventional switching network is given, and derivations of the transfer matrix from a truth table and a structural netlist are provided. To illustrate the use of the vector space model, we describe the exemplary problems of simulation and its inverse, justification, within this framework. One advantage of the vector space information model is that it is a unifying framework with other models in the fields of electrical engineering and computer science. Furthermore, the transformation matrix for a switching network is highly analogous to the concept of a transfer function as found in the areas of linear systems analysis and signal processing since an output response for an information processing network can be obtained through a multiplicative operation between the transformation matrix and an input stimulus modeled as a vector. For this reason, we utilize the term "transfer matrix" interchangeably with the transformation matrix that models a switching network.

Both binary and multi-valued switching networks are modeled, and efficient data structures and methods for manipulation of the underlying vectors and matrices are provided. The transformation of the vector space models into alternative basis is also described. This transformation provides a convenient technique for formulating various spectral transforms and has some advantages when compared to spectral methods employed over switching algebraic models.

The organization of the remainder of the book is as follows. In Chapter 2, we define the notation and basic linear algebraic principles used in the derivations and examples. Chapter 3 contains the definition of the transfer matrix for a switching network given in the form of a truth table and, alternatively, in the form of a structural netlist. The use of the transfer matrix for the purpose of simulation is also included in this chapter. Chapter 4 examines the pseudo-inverse of the transfer matrix and describes a simplification of the pseudo-inverse termed the "justification matrix." The justification matrix is then used for the purpose of obtaining the input stimuli of a switching network given a set of output responses. The results of Chapters 3 and 4 are generalized for the case of multi-valued switching networks in Chapter 5. The subject of spectral representations of switching networks is very naturally developed within the framework of linear algebraic models since the switching network spectrum is conveniently computed through a change of vector space basis. Chapter 6 contains the definitions and methods for computation of various spectra for binary-valued switching networks and Chapter 7 generalizes the results for the case of multi-valued switching networks. For vector space models to be of practical use, it is necessary that they do not incur worse computational complexity than that of traditional Boolean algebraic formulations. Chapter 8 provides a discussion of data structures for representing transfer and justification matrices and also methods for efficiently computing the transfer matrix given various models of conventional switching networks as input such as .pla cube lists and structural net lists expressed in a modern hardware description language such as Verilog or VHDL. Experimental results are also included in Section 8 to illustrate feasibility of the approach. Finally, conclusions and areas of further investigation are provided in Chapter 9.

CHAPTER 2

Information as a Vector

2.1 SWITCHING THEORY

Switching theory is based upon the mathematics of Boole as originally devised for symbolic logic manipulations. The class of Boolean algebras can be defined using a variety of basic operators with a common algebra being defined as $\langle \mathbb{B}, +, \cdot, ^-, 0, 1 \rangle$ where $\mathbb{B} = \{0, 1\}$, $+$ denotes the logical disjunctive or algebraic addition operator (i.e., logical-OR), $\cdot$ denotes the logical conjunctive or algebraic multiplicative operator (i.e., logical-AND), $^-$ denotes the logical complement or algebraic unary inverse operator (i.e., logical-NOT), the value 0 serves as the algebraic additive identity, and the value 1 serves as the algebraic multiplicative identity. This particular algebra is functionally complete with constants. Multi-bit systems can likewise be constructed through the use of the Cartesian product with $\mathbb{B}$ and the resulting algebra is correspondingly functionally complete with constants for such multi-bit systems.

2.2 LINEAR ALGEBRA

A linear algebra is defined over a vector space. Vector spaces are characterized by their dimension and contain elements known as vectors. Vectors are one-dimensional arrays of values or components, and the number of the values comprising a vector defines the vector space dimension in which they are members. The dimension is not to be confused with the tensor order. Tensor order is the number of indices required to specify a tensor. Vectors are tensors of unity order since they require only a single index value to specify the number of components. Likewise, scalars may be viewed as zero-order tensors and matrices as second-order tensors. In general, higher-ordered tensors are definable. The formal definition of a vector space is given in Definition 2.1.

Definition 2.1 *Vector Space*
A vector space consists of a set of vectors and the operations of scaling and addition. □

In general, a vector space may have a dimension approaching infinity although here we only utilize vector spaces with finite dimension. The scaling operation is a multiplicative operation with operands consisting of a scalar and a vector. The product is then the scaled vector that is also a member of the vector space (i.e., closure holds) where each component of the product vector is the scalar multiple of the specified scalar and the original vector component value. The vector addition operation is performed over two operand vectors within the space and the resultant vector sum is comprised of components formed as the scalar sum of corresponding components from the operand vectors. The particular vector spaces we are concerned with are finite-dimensioned Hilbert spaces.

Definition 2.2   *Finite Hilbert Space*
A vector space of dimension k with associated unary norm and binary inner product operations.

$\square$

Column vectors as members of a Hilbert space are denoted as $\mathbf{v} \in \mathbb{H}^k$ and the corresponding row vector as $\mathbf{v}^T$ when the components of $\mathbf{v}$ are not complex-valued. We depart from standard notation slightly in that we specify the dimensionality of the vector space with an integer n where the actual dimension is $k = 2^n$. Thus, we would equivalently denote the $k = 2^n$ dimensional vector $\mathbf{v}$ as $\mathbf{v} \in \mathbb{H}^n$. The reason for this departure from more common notation is to provide correspondence with multi-bit switching systems where 2^n-bit strings are commonly denoted as members of $\mathbb{B}^n$ where $\mathbb{B}^n = \mathbb{B} \times \mathbb{B} \times \ldots \times \mathbb{B}$ and $\times$ denotes the Cartesian product of sets.

A variety of vector norm operations are possible. In this work, we utilize the common Euclidean or L_2 norm defined for a k-dimensional vector $\mathbf{v}$ as

$$L_2(\mathbf{v}) = \|\mathbf{v}\|_2 = \sqrt{\sum_{i=0}^{k-1} v_i^2}$$

where v_i is the ith component of $\mathbf{v}$.

The inner product of two vectors $\mathbf{v}$ and $\mathbf{w}$ is denoted as $\mathbf{v} \cdot \mathbf{w}$ or equivalently as $\mathbf{v}^T \mathbf{w}$ also known as the "dot" product. Mathematically, the inner product among k-dimensional vectors $\mathbf{v}$ and $\mathbf{w}$ is given as

$$\mathbf{v} \cdot \mathbf{w} = \mathbf{v}^T \mathbf{w} = \sum_{i=0}^{k-1} v_i w_i$$

where v_i and w_i are the ith components of $\mathbf{v}$ and $\mathbf{w}$.

A vector $\mathbf{v}$ may undergo a linear transformation that maps it to another vector. Linear transformations are expressed as vector-matrix multiplications where matrices are denoted with capital letters such as $\mathbf{A}$. Matrices are second-order tensors and thus have two defining indices, i, j, where i is the dimensionality of the row space and j the dimensionality of the column space. A matrix can be considered to be a two-dimensional array with indices i and j and can be denoted as $\mathbf{A} = [a_{ij}]$ where a_{ij} is the matrix component in the ith row and j^{th} column. Likewise, matrices are often specified by the maximum value of their indices as the $p \times q$ matrix $\mathbf{A}$ where the row index i satisfies $0 \leq i \leq p - 1$ and the column index j satisfies $0 \leq j \leq q - 1$. We note that another departure from standard notation is that we number index values beginning with zero and ending with $p - 1$ or $q - 1$. This is purposely done to provide for compatibility with Boolean switching functions as will be demonstrated in a later section. A matrix may also be denoted as

$\mathbf{A} = [a_{ij}]_{p \times q}$ when it is desired to explicitly specify the dimensions of the row and column vector space.

The linear transformation of a vector $\mathbf{v} \in \mathbb{H}^n$ resulting in another vector $\mathbf{w} \in \mathbb{H}^n$ is denoted by the direct vector-matrix product operation $\mathbf{Av} = \mathbf{w}$. In this case, $\mathbf{A}$ is square $(i = j)$ and likewise the row vector $\mathbf{w}^T$ can be calculated as $\mathbf{w}^T = \mathbf{v}^T \mathbf{A}^T$ when the components of the vectors and matrices are real-valued. For complex-valued vectors and matrices, $\mathbf{w}^* = \mathbf{v}^* \mathbf{A}^*$ where the superscript $*$ indicates the Hermitian or complex conjugate transpose.

Among the linear transformations within a given vector space $\mathbb{H}^n$, a special case occurs when $\mathbf{Av} = \mathbf{w} = \lambda \mathbf{v}$. When the product vector $\mathbf{w}$ is equivalent to a scaled $\mathbf{v}$, λ is said to be an eigenvalue of $\mathbf{A}$ and $\mathbf{v}$ the corresponding eigenvector of $\mathbf{A}$. Eigenvalues and eigenvectors are only defined for square matrices, that is linear transformations within the same vector space. The eigenvalues λ_i of a transformation matrix $\mathbf{A}$ thus satisfy $\mathbf{Av}_i = \lambda_i \mathbf{v}_i$ which leads to the equivalent equation $(\mathbf{A} - \lambda_i \mathbf{I})\mathbf{v}_i = 0$ where $\mathbf{I}$ is the identity matrix whose diagonal components (i.e., $i = j$) are unity-valued and whose off diagonal components (ie. $i \neq j$) are zero-valued. The eigenvalues can be calculated through the solution of the characteristic equation $|\mathbf{A} - \lambda_i \mathbf{I}| = 0$ where $|\mathbf{A}|$ denotes the determinate of $\mathbf{A}$. The characteristic equation is of the form of an nth-degree polynomial and thus presents several numerical challenges when an algorithmic solution is attempted. Therefore, alternative more numerically stable methods are generally employed rather than explicit solutions of the characteristic polynomial.

The rank of a matrix is an integer that corresponds to the number of column (or row) vectors that are linearly independent. A sufficient condition for linear independence is that the inner product of any of the row (or column) vectors of $\mathbf{A}$ are zero-valued except for the case of a row (or column) vector with itself. Alternatively, it can be shown that a matrix is of full-rank when its eigenvalues are all non-zero and distinct. A special case of a full-rank matrix $\mathbf{A}$ occurs when all row (or column) vectors have $\|\mathbf{v}_i\| = 1$. In this case, the full-rank matrix $\mathbf{A}$ is said to be orthogonal or orthonormal.

The linear transformation relation $\mathbf{Av} = \mathbf{w}$ can be used to solve for a unique $\mathbf{v}$ given $\mathbf{A}$ and $\mathbf{w}$ under certain circumstances only. Those circumstances are that $\mathbf{A}$ must be square and of full rank. In this case, the multiplicative inverse of $\mathbf{A}$ exists and $\mathbf{w} = \mathbf{A}^{-1}\mathbf{v}$. The inverse of a full-rank matrix $\mathbf{A}$ is the solution of the equation $\mathbf{AA}^{-1} = \mathbf{A}^{-1}\mathbf{A} = \mathbf{I}$. As is the case with the computation of eigenvalues, the determination of a matrix inverse can be challenging and a variety of numerical methods are available for this purpose. In the case of a real-valued orthogonal matrix, the inverse is equivalent to the transpose, $\mathbf{A}^{-1} = \mathbf{A}^T$. If the full-rank matrix $\mathbf{A}$ is complex valued and all row (or column) vectors have unity-valued norms, the inverse is equivalent to the Hermitian, $\mathbf{A}^{-1} = \mathbf{A}^*$. An important class of matrices are those where $\mathbf{U}^*\mathbf{U} = \mathbf{UU}^* = \mathbf{I}$. The matrix $\mathbf{U}$ is clearly square and of full rank and furthermore identified as a unitary matrix when this characteristic holds.

2.2.1　VECTOR SPACE MAPPINGS

When vectors are transformed or mapped from an n-dimensional Hilbert space to an m-dimensional Hilbert Space, a non-square transformation matrix $\mathbf{T}$ is used to denote the mapping as $\mathbf{T} : \mathbb{H}^n \rightarrow \mathbb{H}^m$ where $\mathbf{T} = [t_{ij}]_{n \times m}$ and $n \neq m$. Clearly, $\mathbf{T}$ is not of full rank since $n \neq m$, thus the multiplicative inverse does not exist nor does the matrix have eigenvalues.

A characterizing matrix referred to as the square Gram matrix or Gramian is useful for non-square matrices. The square Gram matrix of $\mathbf{T}$ is defined as $\mathbf{G} = \mathbf{A}^T \mathbf{T}$ and eigenvalues exist. In the case of complex-valued matrices, the Gramian is defined as $\mathbf{G} = \mathbf{A}^* \mathbf{A}$. Among other applications, the Gramian is useful for the purpose of computing the singular values of $\mathbf{A}$. In terms of the Gramian $\mathbf{G} = \mathbf{A}^T \mathbf{A}$, the singular values of $\mathbf{A}$ can be calculated as the positive square roots of the eigenvalues of the corresponding Gramian $\mathbf{G}$. The singular values are also useful for a certain type of decomposition known as the "singular value decomposition" and is of the form $\mathbf{T} = \mathbf{U} \mathbf{S} \mathbf{V}^*$ where $\mathbf{U}$ is an $n \times n$ unitary matrix, $\mathbf{S}$ is an $n \times m$ rectangular matrix whose components s_{ii} consist of the singular values, and $\mathbf{V}$ is an $m \times m$ unitary matrix.

2.2.2　BRA-KET NOTATION AND THE OUTER PRODUCT

To provide compatibility with the quantum logic and computing community, we utilize the "bra-ket" notation of Paul Dirac [5]. Column vectors are referred to as 'kets' and denoted $|v\rangle$ which is identical to the more standard boldface font notation of $\mathbf{v}$. Likewise, a row vector $\mathbf{v}^T$ is referred to as "bra-v" and denoted $\langle v|$. The inner product $\mathbf{v}^T \mathbf{w}$ is written as $\langle v|w \rangle$. The relationship between the norm of a vector and the inner product of the vector with itself is $\mathbf{v} \cdot \mathbf{v} = \mathbf{v}^T \mathbf{v} = [L_2(\mathbf{v})]^2 = [\|\mathbf{v}\|]^2 = \langle v|v \rangle$.

The outer product is a binary multiplicative operation that can be used to multiply two tensors regardless of their respective order. For the finite-dimensioned Hilbert spaces we are concerned with here, the outer product is identical to the Kronecker product of matrices. The outer product of two vectors $\mathbf{v}$ and $\mathbf{w}$ is denoted as $\mathbf{v} \otimes \mathbf{w} = \mathbf{u}$ where the dimension of $\mathbf{u}$ is the sum of the dimensions of $\mathbf{v}$ and $\mathbf{w}$. The use of bra-ket notation is very convenient in this work since inner products are expressed as $\langle v|w \rangle$ while the outer product is $|v\rangle\langle w|$. Mathematically the outer product of an n-dimensional vector $\mathbf{v}$ with an m-dimensional vector $\mathbf{w}$ is expressed as

$$|v\rangle\langle w| = \begin{bmatrix} v_0 w_0 \\ v_0 w_1 \\ \vdots \\ v_0 w_{m-1} \\ v_1 w_0 \\ v_1 w_1 \\ \vdots \\ v_1 w_{m-1} \\ \vdots \\ v_{n-1} w_0 \\ v_{n-1} w_1 \\ \vdots \\ v_{n-1} w_{m-1} \end{bmatrix}.$$

Note that the outer product operation is not commutative since $|v\rangle\langle w| \neq |w\rangle\langle v|$. The outer product is useful for representing several vectors in lower-dimensioned Hilbert spaces as a single vector in a higher-dimensioned Hilbert space. When $|v\rangle \in \mathbb{H}^n$ and $|w\rangle \in \mathbb{H}^m$, the outer product $|v\rangle\langle w| \in \mathbb{H}^{n+m}$. Bra-ket notation is convenient since the orientation of the respective bra or ket within a mathematical expression implicitly indicates whether a multiplication is carried out using the inner or outer product. The expression $\mathbf{A}|v\rangle$ indicates that a direct vector-matrix product is to be carried out resulting in a product vector $|w\rangle$ of the same dimension as that of $|v\rangle$ and furthermore $\mathbf{A}|v\rangle = \langle v|\mathbf{A}^T$ for real-valued $\mathbf{A}$. The expression $|v\rangle\mathbf{A}$ where $|v\rangle$ is of dimension n and $\mathbf{A}$ of dimension $m \times p$ would denote the outer product of $\mathbf{v} \otimes \mathbf{A}$ and result in a matrix of dimension $(nm) \times p$.

2.3   MODELING INFORMATION IN THE HILBERT SPACE

Binary switching variables x_i can be expressed as $x_i = m_0 \cdot 0 + m_1 \cdot 1$ where $m_i \in \mathbb{B}$. The variable represents an atomic information datum in the form of a binary digit or "bit" and has value 1 when $m_0 = 0$ and $m_1 = 1$, and likewise has value 0 when $m_0 = 1$ and $m_1 = 0$. indexbit This allows for a convenient definition of the negation operation denoted as $\bar{x}$ by interchanging the coefficients m_i as $\bar{x} = m_1 \cdot 0 + m_0 \cdot 1$. The corresponding linear algebraic model for atomic binary information data values is given in Definition 2.3.

Definition 2.3   *Linear Algebraic Model for Binary Information Values*
Atomic information values are modeled as the basis vectors $\langle 0|$ and $\langle 1|$ where $\langle 0| = [1 \ \ 0]$ and $\langle 1| = [0 \ \ 1]$. $\langle 0|$ corresponds to the switching algebra bit valued as 0 while $\langle 1|$ corresponds to the switching algebra bit valued as 1.   □

Just as multi-bit input vectors can be expressed as elements of $\mathbb{B}^n$, a corresponding vector $\langle x| \in \mathbb{H}^n$ can be formulated from the individual binary values $\langle x_i| \in \mathbb{H}$ through the use of the outer product. Example 2.4 illustrates the use of the outer product to represent a single input stimulus vector for a switching network with n primary inputs.

Example 2.4   *Single Input Stimulus Vector for n-input Network*
Consider the case where a 3-input switching network is excited with an input stimulus whose values are $x_0 = \langle 1|$, $x_1 = \langle 0|$, and $x_2 = \langle 1|$. In this case each primary input value is modeled as a vector $\langle x_i| \in \mathbb{H}$. Because the outer product can be used to expand the dimensionality of a Hilbert space, a single 2^3-dimensioned vector representing the input stimulus becomes

$$\langle 101| = \langle 1| \otimes \langle 0| \otimes \langle 1| = \begin{bmatrix} 0 & 1 \end{bmatrix} \otimes \begin{bmatrix} 1 & 0 \end{bmatrix} \otimes \begin{bmatrix} 0 & 1 \end{bmatrix}$$
$$= \begin{bmatrix} 0 & 0 & 0 & 0 & 0 & 1 & 0 & 0 \end{bmatrix}$$

$\square$

Definition 2.5   *Canonical Basis Vector*
Canonical basis vectors are defined as those vectors whose components are all zero-valued except for a single unity-valued component. $\square$

Observation 2.6   *Specific Information Valuations are Canonical Basis Vectors*
The specific valuation of a primary input stimulus vector or primary output response vector for a switching network is a canonical basis vector. $\square$

Definition 2.6 and Example 2.4 infer some useful insight with regard to decomposing vectors in higher-dimensioned Hilbert spaces into a series of outer product factors. Since the vectors representing single data values are canonical basis vectors, it is often useful to express the vector in an alternative number system base or radix and then to determine the index of the unity-valued component. When not implicitly clear, a value may be written in parentheses with a subscript (in the decimal number system) that is the number system radix. For example, $\langle (6)_{10}| = \langle (110)_2| = \langle (20)_3|$. Example 2.7 shows how the change of base for a canonical basis vector is useful in determining the outer product factors.

Example 2.7   *Factoring Canonical Basis Vectors*
Consider the canonic basis vector $\langle (1011)_2|$. The basis vector as written in array notation can be computed using the Hilbert space expansion relation as $\langle (1011)_2| = \langle 1| \otimes \langle 0| \otimes \langle 1| \otimes \langle 1|$ and this calculation would yield

$$\langle 1011| = \begin{bmatrix} 0 & 1 \end{bmatrix} \otimes \begin{bmatrix} 1 & 0 \end{bmatrix} \otimes \begin{bmatrix} 0 & 1 \end{bmatrix} \otimes \begin{bmatrix} 0 & 1 \end{bmatrix}$$
$$= \begin{bmatrix} 0 & 0 & 0 & 0 & 0 & 0 & 0 & 0 & 0 & 0 & 0 & 1 & 0 & 0 & 0 & 0 \end{bmatrix}.$$

However, using a simple change of base, we see that $\langle (1011)_2 | = \langle (11)_{10} |$. Thus, the corresponding row vector when explicitly written has zero-valued components everywhere except the 11th unity-valued component.

For various EDA applications, it is necessary to define special additional values in conventional Boolean switching algebras such as the use of a "don't care" often denoted by X, or as in the case of hardware description languages (HDL), an indeterminate value which is also unfortunately denoted as X. Other common values used in HDLs and other EDA environments include high-impedance usually denoted as Z and several others [16] [17]. The vector space model proposed here allows for such values to be conveniently modeled with natural extensions to the form of the vector representing an information quantity.

The concept of an "input don't care" versus internal and output don't cares has been described in [13]. While these are characterized as different types of don't cares, they really have quite different meanings. The input don't care is often denoted with a – in cube list formats such as the .pla format. For example, the input or AND-plane cube denoted as 0-1 in the .pla format indicates that both 001 and 011 are covered. This means that – indicates the middle primary input has both binary digit values of 0 and 1. In the vector space model, this is denoted as the "total vector" $\langle t |$ where $\langle t | = \langle 0 | + \langle 1 | = [0\ \ 1] + [1\ \ 0] = [1\ \ 1]$. Another useful quantity is the null vector indicating that the datum has no value; it is neither $\langle 0 |$ nor $\langle 1 |$ and is denoted as $\langle \varnothing |$. The use of $\langle t |$ and $\langle \varnothing |$ allows for a complete lattice to be formulated with $\langle t |$ serving as the greatest lower bound (glb) since it covers all other values and $\langle \varnothing |$ as the least upper bound (lub) since it covers no other values.

When a true third value is required, such as that of high-impedance, the dimension of the Hilbert space may be increased. As an example, if a switching algebra quantity can take on values from the set $\{0, 1, Z\}$, the corresponding vector space model could be $\langle 0 | = [1\ \ 0\ \ 0]$, $\langle 1 | = [0\ \ 1\ \ 0]$, and $\langle Z | = [0\ \ 0\ \ 1]$. As is more fully explained in a later chapter, values such as $\langle Z |$ are very useful for modeling switching networks comprised of components such as tri-state buffers or other three-state circuits.

CHAPTER 3

Switching Network Transfer Functions

3.1 SWITCHING NETWORK MODELS

Switching networks are conventionally modeled using a set of Boolean switching functions where each of the m primary network outputs are modeled with separate switching functions. We propose to model such networks as the transformation of a vector $\langle x| \in \mathbb{H}^n$ representing the network input stimulus to a corresponding vector $\langle f| \in \mathbb{H}^m$ and the specific functionality of the network with a transformation matrix $\mathbf{F}$. Thus a switching network is modeled as the mathematical mapping of a vector from one Hilbert space to another, $\mathbf{F} : \mathbb{H}^n \rightarrow \mathbb{H}^m$. Figure 3.1 contains a conceptual diagram of a switching network modeled conventionally with Boolean switching algebra on the left and with the proposed vector space model on the right.

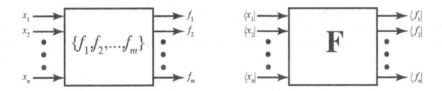

Figure 3.1: Conceptual models of switching networks.

The set of switching functions $\{f_1, f_2, \ldots, f_m\}$ are symbolically denoted using operators from a Boolean algebra such as $\langle \mathbb{B}, +, \cdot, ^-, 0, 1 \rangle$ where the atomic network elements or "logic gates" correspond to algebraic operators or expressions involving the operators. For example, an AND gate corresponds to the multiplicative operator denoted by $\cdot$ and the exclusive-OR gate corresponds to the expression $x \cdot \bar{y} + \bar{x} \cdot y$. The primary inputs x_i are switching variables over the set $\mathbb{B}$ as are the individual network primary outputs denoted as f_i. A specific input stimulus vector may be denoted as an n-dimensional vector whose components are each $x_i \in \mathbb{B}$, or alternatively, as a single element from $\mathbb{B}^n$. Likewise, the output response vector can be considered to be an element of $\mathbb{B}^m$.

The proposed model utilizes a vector space transformation with a corresponding linear algebra. Definition 3.1 defines the proposed algebra.

Definition 3.1   *Linear Algebraic Model for Switching Networks*
The algebra denoted as $\langle \mathbb{H}, +, \cdot, \langle 0|, \langle 1| \rangle$ is used to model the functionality of a switching network where $+$ denotes vector addition, $\cdot$ represents the inner product, and $\langle 0|$ and $\langle 1|$ represent the canonical basis vectors of $\mathbb{H}$.   □

3.2   SWITCHING NETWORK TRANSFER MATRIX DERIVATION

The objective of this section is to derive the transformation matrix **F** appearing in the right-hand portion of Fig. 3.1. **F** characterizes a particular switching network and is a single transformation matrix that projects network input stimulus vectors $\langle x_i| \in \mathbb{H}^n$ to corresponding network output response vectors $\langle f_i| \in \mathbb{H}^m$, or, more concisely $\mathbf{F} : \mathbb{H}^n \to \mathbb{H}^m$. For the purpose of consistency with the literature in classical linear systems theory [14] and in the quantum logic community [15], we interchangeably refer to the transformation matrix **F** as a "transfer matrix" since it can be used to obtain the output response of a network by a multiplicative operation with a corresponding input stimulus and is thus analogous to a transfer function as commonly used in classical linear systems analysis.

We shall first consider the derivation of a projection matrix $\mathbf{P}_i$ that projects a single vector $\langle x_i| \in \mathbb{H}^n$ to a corresponding vector $\langle f_i| \in \mathbb{H}^m$, or, equivalently, $\mathbf{P}_i : \mathbb{H}^n \to \mathbb{H}^m$. The projection is in the form of a linear transformation, thus, $\langle x_i|\mathbf{P}_i = \langle f_i|$. Taking the outer product of each side of this equation by multiplying with $|x_i\rangle$ yields Equation 3.1.

$$|x_i\rangle\langle x_i|\mathbf{P}_i = |x_i\rangle\langle f_i| \tag{3.1}$$

The matrix $\mathbf{P}_i$ is of dimension $n \times m$ and is, in general, non square, and when it does happen to be square, it is often not of full rank. Thus $|x_i\rangle\langle x_i|^{-1}$ does not usually exist and cannot be formed to solve Equation 3.1. However, the form of $|x_i\rangle\langle x_i|$ is that of a "Dirac-delta matrix" due to Observation 2.6, $\langle x_i|$ is a canonical basis vector since it represents the ith valuation of an input stimulus vector.

Definition 3.2   *Dirac-delta Matrix*
A Dirac-delta matrix is denoted as $\delta_{pq} = [\delta_{ij}]_{n \times m}$ contains a single unity-valued component defined by a set of indices, pq. The components of the Dirac-delta matrix δ_{ij} are given by the scalar Dirac-delta function

$$\delta_{ij} = \begin{cases} 1, & i = p, j = q \\ 0, & otherwise \end{cases}.$$

$\square$

Using Definition 3.2, Equation 3.1 can be rewritten as

$$\delta_{pq} \mathbf{P}_i = |x_i\rangle\langle f_i|.$$

Because the row-space of δ_{pq} consists of $n - 1$ null row vectors and the pth row vector is a canonical basis vector $\langle q|$, $n - 1$ row vectors of $\mathbf{P}_i$ have infinitely many solutions including the solution of setting them equal to the null vector $\langle \varnothing| \in \mathbb{H}^m$. The qth row vector of $\mathbf{P}_i$ is not null and is denoted $\langle p_q|$. Likewise, the matrix $|x_i\rangle\langle f_i|$ consists of $n - 1$ null row vectors with the qth row vector equal to $\langle f_i|$.

Using these observations, we choose the solution of allowing the $n - 1$ row vectors of $\mathbf{P}_i$ to be $\langle \varnothing|$ and we rewrite Equation 3.1 as Equation 3.2.

$$\langle p_q| = \langle f_i| \tag{3.2}$$

The preceding derivation allows Lemma 3.3 to be stated.

Lemma 3.3 i^{th} Row Vector of $\mathbf{P}_i$
A projection matrix $\mathbf{P}_i$ for a logic network characterized by a transfer matrix $\mathbf{F}$ may be written as an $n \times m$ matrix with $n - 1$ null row vectors $\langle \varnothing| \in \mathbb{H}^m$ and with the ith row vector equivalent to the ith valuation of the logic network $\langle f_i|$ as

$$\langle p_i| = \langle f_i|.$$

Proof. The derivations preceding Lemma 3.3 serve as a proof. $\square$

In determining the form of the overall transfer matrix $\mathbf{F}$ that characterizes a switching network, it is necessary to formulate the matrix such that the input stimulus $\langle x_i|$ when multiplied with $\mathbf{F}$ results in the ith valuation of the switching network. Lemma 3.3 proves that the ith row vector of $\mathbf{P}_i$ is identically equal to the output response of logic network.

Observation 3.4 *Complete Input Stimulus Matrix is* $\mathbf{I}$
Because each specific input stimulus vector for a logic network is of the form of a canonical basis vector, the collection of all input stimuli vectors form a complete canonical basis of $\mathbb{H}^n$. The matrix

$\mathbf{X}$ is formed as the set of all possible input stimuli as row vectors ordered from $i = 0$ to $i = n - 1$ from top to bottom and is equivalent to the identity matrix $\mathbf{I}$.

Theorem 3.5   Transfer Matrix of a Logic Network
The transfer matrix $\mathbf{F}$ characterizing a logic network with n primary inputs and m primary outputs is expressed as shown in Equation 3.3.

$$\mathbf{F} = \sum_{i=0}^{2^n-1} \mathbf{P}_i \tag{3.3}$$

Proof. A solution for a form of $\mathbf{P}_i$ is that it consists of $n - 1$ null row vectors with the ith row vector equivalent to $\langle f_i |$ as given in Lemma 3.3. Because the set of input stimulus vectors are linearly independent basis vectors $\langle x_i | = \langle i |$, the collection of all projection matrices $\mathbf{P}_i$ contain row vectors equivalent to $\langle f_i |$ distributed among all differing row indices i. Therefore,

$$\mathbf{XF} = \sum_{i=0}^{2^n-1} \mathbf{P}_i. \tag{3.4}$$

From Observation 3.4, $\mathbf{X} = \mathbf{I}$, substituting this result in Equation 3.4 yields Equation 3.3.
□

Using the result of Equation 3.1 and Observation 3.4 with the result of Theorem 3.3, Corollary 3.6 results allowing the transfer matrix of a logic network to be formulated as a sum of outer products.

Corollary 3.6   Transfer Matrix as Sum of Outer Products
The transfer matrix for a logic network characterized by $\mathbf{F}$ can be formed as a sum of outer products as given in Equation 3.5.

$$\mathbf{F} = \sum_{i=0}^{2^n-1} |i\rangle\langle f_i| \tag{3.5}$$

Proof. From Equation 3.1 and the fact that input stimulus vectors take the form of canonical basis vectors, we can obtain the expression:

$$|i\rangle\langle i|\mathbf{P}_i = |i\rangle\langle f_i|.$$

$|i\rangle\langle i|$ is a square Dirac-delta matrix whose i^{th} row vector is $\langle i|$ allowing for the previously noted solution of $\mathbf{P}_i$ to be defined as consisting of null row vectors for all but the ith row vector which is equivalent to $\langle f_i|$. Therefore, the projection matrix $\mathbf{P}_i = |i\rangle\langle f_i|$. Substituting this result into the result of Theorem 3.5 yields Equation 3.5.

3.3   TRANSFER MATRICES OF COMMON SWITCHING CIRCUITS

Equation 3.3 may be used to determine the transfer matrices for common switching network components serving as atomic operators or, "logic gates." Example 3.7 illustrates the calculation of the transfer matrix $\mathbf{A}$ for a two-input AND gate.

Example 3.7   *Transfer Matrix for 2-input AND Gate*
Consider a 2-input AND gate whose symbol and switching algebra truth table are shown in Figure 3.2. The four possible input stimuli are $\langle 00|$, $\langle 01|$, $\langle 10|$, and $\langle 11|$. The projection matrix relationships for the AND gate then become $\langle 00|\mathbf{P}_0 = \langle 0|$, $\langle 01|\mathbf{P}_1 = \langle 0|$, $\langle 10|\mathbf{P}_2 = \langle 0|$, and $\langle 11|\mathbf{P}_3 = \langle 1|$. From Lemma 3.3, the projection matrices are of the form:

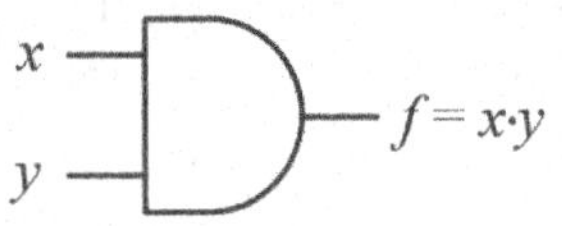

x	y	$f = x \cdot y$
0	0	0
0	1	0
1	0	0
1	1	1

Figure 3.2: Circuit symbol and truth table for binary AND gate.

$$\mathbf{P}_0 = \begin{bmatrix} 1 & 0 \\ 0 & 0 \\ 0 & 0 \\ 0 & 0 \end{bmatrix} \quad \mathbf{P}_1 = \begin{bmatrix} 0 & 0 \\ 1 & 0 \\ 0 & 0 \\ 0 & 0 \end{bmatrix} \quad \mathbf{P}_2 = \begin{bmatrix} 0 & 0 \\ 0 & 0 \\ 1 & 0 \\ 0 & 0 \end{bmatrix} \quad \mathbf{P}_3 = \begin{bmatrix} 0 & 0 \\ 0 & 0 \\ 0 & 0 \\ 0 & 1 \end{bmatrix}$$

The overall transfer matrix for a two-input AND gate $\mathbf{A}$ is given by Equation 3.3.

$$A = \sum_{i=0}^{3} P_i = P_0 + P_1 + P_2 + P_3$$

$$= \begin{bmatrix} 1 & 0 \\ 0 & 0 \\ 0 & 0 \\ 0 & 0 \end{bmatrix} + \begin{bmatrix} 0 & 0 \\ 1 & 0 \\ 0 & 0 \\ 0 & 0 \end{bmatrix} + \begin{bmatrix} 0 & 0 \\ 0 & 0 \\ 1 & 0 \\ 0 & 0 \end{bmatrix} + \begin{bmatrix} 0 & 0 \\ 0 & 0 \\ 0 & 0 \\ 0 & 1 \end{bmatrix} = \begin{bmatrix} 1 & 0 \\ 1 & 0 \\ 1 & 0 \\ 0 & 1 \end{bmatrix}$$

□

This calculation can be carried out for any logic network; however, it is clearly exponentially complex as the number of network inputs increase. However, for small two- and three-input sub circuits and structures, the calculation can be easily performed. In a following section, we show that the transfer matrix for a larger network can be hierarchically formulated using smaller transfer matrices and thus the exponential complexity present in Equation 3.3 can be avoided. Three important logic network topological configurations can occur, and transfer matrices are required for these instances as well as a library of logic gate matrices to employ the hierarchical methodology.

3.3.1   FANOUT STRUCTURE TRANSFER MATRIX

A fanout is an electrical circuit node in which a single conducting wire carries a signal that drives two or more conductors. Figure 3.3 shows the circuit diagram topology for a fanout of size two in the leftmost diagram. From a switching circuit point of view, this is analogous to copying an input datum to two or more resultant data. Electrically, there are clearly limitations to fanout, but a discussion of this topic is beyond the scope of this book. Fanout or the copying of information values is forbidden in quantum logic due to the no-cloning theorem; however, fanout is permissible and is responsible for circuit area reduction in conventional electronic switching circuits.

The transfer matrix $\mathbf{F_O}$ for a two-output fanout is computed using Equation 3.5 as:

$$\mathbf{F_O} = |0\rangle\langle 00| + |1\rangle\langle 11| = \begin{bmatrix} 0 \\ 1 \end{bmatrix} \begin{bmatrix} 1 & 0 & 0 & 0 \end{bmatrix} + \begin{bmatrix} 1 \\ 0 \end{bmatrix} \begin{bmatrix} 0 & 0 & 0 & 1 \end{bmatrix}$$

$$= \begin{bmatrix} 1 & 0 & 0 & 0 \\ 0 & 0 & 0 & 1 \end{bmatrix}$$

3.3.2   FANOUT STRUCTURE TRANSFER MATRIX

The fanin structure is depicted as the diagram in the center of Figure 3.3. The fanin structure is only used in special cases such as when the implementation technologies enables "wired-logic"

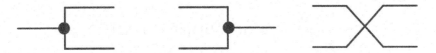

Figure 3.3: Circuit diagram structures for fanout, fanin, and crossover.

switching operators. Certain electronic technologies allow for fanin structures due to other circuit element outputs being at a high-impedance or other state. As an example, certain current-mode circuitry allow for a fanin to operate as an AND-type operation. In those cases, the fanin transfer matrix is identical to that of the matrix characterizing the appropriate switching circuit gate. Many voltage-mode circuits can only include fanin structures when the inputs are driven by disjoint outputs thereby avoiding a short-circuit situation. In general, fanins are not allowed and thus we model them in this case with null row vectors for the disallowed case. Null vectors occur in the transfer matrix when those specific disallowed input cases are simply excluded in the calculation of the matrix. For these reasons, the fanin transfer matrix is technology dependent.

In the case where fanins with differing input stimulus values are disallowed, the transfer matrix may be computed using the relationship in Equation 3.5 as:

$$\mathbf{F_I} = |00\rangle\langle0| + |11\rangle\langle1| = \begin{bmatrix} 1 \\ 0 \\ 0 \\ 0 \end{bmatrix} \begin{bmatrix} 1 & 0 \end{bmatrix} + \begin{bmatrix} 0 \\ 0 \\ 0 \\ 1 \end{bmatrix} \begin{bmatrix} 0 & 1 \end{bmatrix} = \begin{bmatrix} 1 & 0 \\ 0 & 0 \\ 0 & 0 \\ 0 & 1 \end{bmatrix}$$

3.3.3   CROSSOVER STRUCTURE TRANSFER MATRIX

In a later section, we will describe how the transfer matrix can be constructed for an interconnection of basic operators, or, a switching network. Because the network transfer matrix is dependent upon the topology of the network, the case where conductors cross one another in the plane must be accounted for. Multiple crossovers can be dealt with as a series of single crossovers, thus a fundamental structure to be considered is the single crossing of two conductors that are electrically isolated. Such a structure is depicted as the rightmost diagram in Figure 3.5. The crossover matrix expresses the four input-output relationships $\langle00| \to \langle00|$, $\langle01| \to \langle10|$, $\langle10| \to \langle01|$, and $\langle11| \to \langle11|$. Equation 3.5 indicates that the transfer matrix can be computed as:

$$\mathbf{C} = |00\rangle\langle00| + |01\rangle\langle10| + |10\rangle\langle01| + |11\rangle\langle11|$$

$$= \begin{bmatrix} 1 \\ 0 \\ 0 \\ 0 \end{bmatrix} \begin{bmatrix} 1 & 0 & 0 & 0 \end{bmatrix} + \begin{bmatrix} 0 \\ 1 \\ 0 \\ 0 \end{bmatrix} \begin{bmatrix} 0 & 0 & 1 & 0 \end{bmatrix}$$

$$+ \begin{bmatrix} 0 \\ 0 \\ 1 \\ 0 \end{bmatrix} \begin{bmatrix} 0 & 1 & 0 & 0 \end{bmatrix} + \begin{bmatrix} 0 \\ 0 \\ 0 \\ 1 \end{bmatrix} \begin{bmatrix} 0 & 0 & 0 & 1 \end{bmatrix}$$

$$= \begin{bmatrix} 1 & 0 & 0 & 0 \\ 0 & 0 & 1 & 0 \\ 0 & 1 & 0 & 0 \\ 0 & 0 & 0 & 1 \end{bmatrix}$$

3.3.4   OTHER BASIC SWITCHING ELEMENTS

Other basic switching circuit elements can be computed in a similar manner to that shown in Example 3.7 for the AND gate. Two cases deserving of mention are the non-inverting buffer and the single conductor or pass-through line. In both of these cases, the appropriate transfer matrix is the 2×2 identity matrix. Although the identity matrix causes the logic network model to produce non-transformed outputs, one cannot assume that non-inverting buffers can be replaced with single pass-through conductors. Non-inverting buffers are often included to perform slack-matching timing operations or to restore the voltage or current levels to acceptable values. The abstraction level of the switching models described here do not include such information regarding timing and voltage or current levels, thus the transfer matrices remain identical and are shown on the right-hand side of Figure 3.5 along with several other common basic switching network elements and their respective transfer matrices.

It is noted that the gates with negated outputs are related to their positive-polarity output counterparts through the relationship

$$\mathbf{N}g = [\mathbf{1}] - g$$

where g indicates the gate type and $[\mathbf{1}]$ denotes the matrix whose components are all unity-valued.

3.4   LOGIC NETWORK TRANSFER MATRICES

Logic networks consist of interconnections of basic elements such as the example depicted in Figure 3.4. In general, transfer matrices that model logic networks can be computed using the

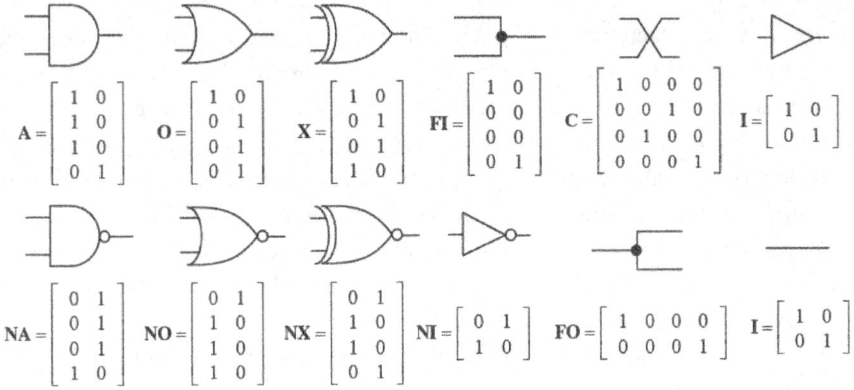

$$A = \begin{bmatrix} 1 & 0 \\ 1 & 0 \\ 1 & 0 \\ 0 & 1 \end{bmatrix} \quad O = \begin{bmatrix} 1 & 0 \\ 0 & 1 \\ 0 & 1 \\ 0 & 1 \end{bmatrix} \quad X = \begin{bmatrix} 1 & 0 \\ 0 & 1 \\ 0 & 1 \\ 1 & 0 \end{bmatrix} \quad FI = \begin{bmatrix} 1 & 0 \\ 0 & 0 \\ 0 & 0 \\ 0 & 1 \end{bmatrix} \quad C = \begin{bmatrix} 1 & 0 & 0 & 0 \\ 0 & 0 & 1 & 0 \\ 0 & 1 & 0 & 0 \\ 0 & 0 & 0 & 1 \end{bmatrix} \quad I = \begin{bmatrix} 1 & 0 \\ 0 & 1 \end{bmatrix}$$

$$NA = \begin{bmatrix} 0 & 1 \\ 0 & 1 \\ 0 & 1 \\ 1 & 0 \end{bmatrix} \quad NO = \begin{bmatrix} 0 & 1 \\ 1 & 0 \\ 1 & 0 \\ 1 & 0 \end{bmatrix} \quad NX = \begin{bmatrix} 0 & 1 \\ 1 & 0 \\ 1 & 0 \\ 0 & 1 \end{bmatrix} \quad NI = \begin{bmatrix} 0 & 1 \\ 1 & 0 \end{bmatrix} \quad FO = \begin{bmatrix} 1 & 0 & 0 & 0 \\ 0 & 0 & 0 & 1 \end{bmatrix} \quad I = \begin{bmatrix} 1 & 0 \\ 0 & 1 \end{bmatrix}$$

Figure 3.4: Basic logic network elements and transfer matrices.

relationship in Equation 3.5. Clearly, this method is impractical for larger networks since it requires the formulation and summation of an exponentially large number of outer product terms (i.e., exponential with respect to the number of primary inputs, n).

A more efficient method for determination of a transfer matrix for a logic network allows the matrix to be formed through a traversal of a representation of the topological structure of the network. Such representations are commonly represented as structural HDL descriptions. Example 3.8 depicts a simple logic network in the form of a Verilog structural description and an accompanying circuit diagram. Such structural HDL forms are textual representations of the corresponding circuit diagrams and it is a common task for modern EDA algorithms to parse the HDL into an internal data structure representing the topological circuit diagram.

Example 3.8   *Logic Network HDL and Circuit Diagram*
A two-input, two-output logic network is depicted in the form of a Verilog structural netlist and as a circuit diagram in Figure 3.5.

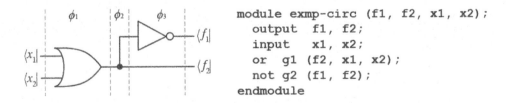

```
module exmp-circ (f1, f2, x1, x2);
    output  f1, f2;
    input   x1, x2;
    or  g1 (f2, x1, x2);
    not g2 (f1, f2);
endmodule
```

Figure 3.5: Example logic network HDL and diagram.

The circuit diagram in Figure 3.8 is annotated with three vertical dashed lines depicting serial partitions. Within each partitions, the logic network elements are in parallel and thus the lines crossing each partition line each carry individually distinct logic values corresponding to a particular input stimulus vector. These individual logic vectors are elements of $\mathbb{H}^2$ and may be combined into a single vector in a higher-dimensioned vector space by multiplying them with the outer product operation. The outer product operation is non-commutative, thus the order in which the multiplications are applied must be consistent for each partition. In the work presented here, we arbitrarily choose to use the topmost value as the leftmost factor in the outer product calculation. Thus, the input stimulus vector is represented as $\langle x_1 x_2| = \langle x_1| \otimes \langle x_2|$. Likewise, the output response vector is $\langle f_1 f_2| = \langle f_1| \otimes \langle f_2|$.

The partitions are denoted as ϕ_1, ϕ_2, and ϕ_3. The vector describing the logic value at partition ϕ_1 results from a transformation of $\langle x_1 x_2| \in \mathbb{H}^2$ to a corresponding vector $\langle w_1| \in \mathbb{H}$ at the output net of the AND gate. $\langle w_1|$ can thus be obtained through a direct vector-matrix product operation $\langle x_1 x_2|\mathbf{A} = \langle w_1|$. Likewise, the values at partition ϕ_2 result from $\langle w_1|\mathbf{FO} = \langle w_2 w_3|$. Partition ϕ_3 consists of two parallel circuit elements, an inverter (or NOT gate) near the top of the figure and a pass-through element near the bottom resulting in $\langle f_1|$ and $\langle f_2|$. The output response may be formed as a single vector through the outer product operation $\langle f_1| \otimes \langle f_2| = \langle f_1 f_2|$. Since $\langle f_1| = \langle w_2|\mathbf{N}$ and $\langle f_2| = \langle w_3|\mathbf{I}$, these two relationships may be combined using the outer product yielding $\langle f_1 f_2| = (\mathbf{N} \otimes \mathbf{I})\langle w_2 w_3|$. This observation allows the overall transformation matrix for a particular petition to be formed as the outer product of each of the parallel elements comprising the partition. Furthermore, the overall partition matrices of each partition can be combined using the direct matrix product due to the multiplicative relationship among the input stimuli and output responses using the transfer matrix. These observations allow the overall transfer matrix $\mathbf{T}$ for the logic network in Figure 3.5 to be computed in symbolic factored form as:

$$\mathbf{T} = (\mathbf{A})(\mathbf{FI})(\mathbf{N} \otimes \mathbf{I}).$$

This technique of extracting a transfer matrix directly from a structural netlist is important since the method allows for the matrix to be formed without resorting to forming an exponential number of matrices to be summed together as described in the relationship in Equation 3.5. The overall complexity of the computation becomes $O(N)$ where N is the number of basic logic elements. The explicit representation of the transfer matrix for the circuit in Figure 3.5 is given in Equation 3.6.

$$\mathbf{T} = \begin{bmatrix} 1 & 0 \\ 1 & 0 \\ 1 & 0 \\ 0 & 1 \end{bmatrix} \begin{bmatrix} 1 & 0 & 0 & 0 \\ 0 & 0 & 0 & 1 \end{bmatrix} \left(\begin{bmatrix} 0 & 1 \\ 1 & 0 \end{bmatrix} \otimes \begin{bmatrix} 1 & 0 \\ 0 & 1 \end{bmatrix} \right) = \begin{bmatrix} 0 & 1 & 0 & 0 \\ 0 & 0 & 1 & 0 \\ 0 & 0 & 1 & 0 \\ 0 & 0 & 1 & 0 \end{bmatrix} \tag{3.6}$$

While the netlist partitioning and traversal method reduces the number of computations required to determine a logic network transfer matrix, the overall matrix remains exponentially large. Fortunately, efficient data structures exist for representing the transfer matrices that, on average, allow them to require greatly reduced storage. A discussion of efficient storage methods is provided in a following chapter and example computational results are provided to evaluate the effectiveness of this method.

CHAPTER 4

Simulation and Justification

4.1 NETWORK OUTPUT RESPONSE

The transfer matrix characterizes a switching network and provides a method for simulation: the determination of a network output response given an input stimulus through a multiplicative operation. Simulation is a very commonly employed task in logic network design and analysis and is a core EDA capability. Justification is the inverse operation where an input stimulus is determined given a network transfer matrix and output response. Justification has applications in the areas of test set generation, reverse engineering, and validation approaches.

Theorem 4.2 proves that a switching network output response is obtained through a multiplicative operation among the transfer matrix and input stimulus vector.

Lemma 4.1 Independence of Network Input Vectors
The collection of vectors representing a specific valuation of switching network primary input assignments is linearly independent.

Proof. Consider two vectors $\langle x_i|$ and $\langle x_j|$ representing two specific valuations of the n primary inputs of a switching network. A specific valuation is the assignment of either $\langle 0|$ or $\langle 1|$ to each primary input. Expressing the input stimulus vectors as elements of $\mathbb{H}^n$, they are written as $\langle (b_0 b_1 \ldots b(n-1))_2|$ and from Observation 2.6, are in the form of canonic basis vectors over $\mathbb{H}^n$. Canonic basis vectors have a unity-valued norm since $\langle x_i|x_i\rangle = 1$ and form inner products of the form:

$$\langle x_i|x_j\rangle = \begin{cases} 0, & i \neq j \\ 1, & i = j \end{cases}$$

By definition, two vectors with a non-zero norm and whose inner product is zero are linearly independent. $\square$

Theorem 4.2 Switching Network Output Response
The output response of a switching network $\langle f_i|$ is obtained as the direct vector-matrix product of the characterizing transfer matrix $\mathbf{F}$ and an input stimulus vector $\langle x_i|$ as expressed in Equation 4.1.

$$\langle f_i| = \langle x_i|\mathbf{F} \tag{4.1}$$

Proof. The transfer matrix **F** characterizing a logic network is of the form of Equation 3.5. Thus, multiplying this expression with an input stimulus vector $\langle x_i|$ yields:

$$\langle x_i|\mathbf{F} = \langle x_i| \sum_{j=0}^{2^n-1} |x_j\rangle\langle f_j| = \sum_{j=0}^{2^n-1} \langle x_i|x_j\rangle\langle f_j|$$

Using the result of Lemma 4.1, the summation argument is zero-valued for all but the case where $i = j$, and the result becomes $\langle f_i| = \langle x_i|\mathbf{F}$. $\qquad\square$

Network Response Calculation using the Monolithic Transfer Matrix
Multiple output responses may be obtained with a single evaluation of Equation 4.1 by assigning the value $\langle t| = \langle 0| + \langle 1|$ to one or more switching network primary inputs. Furthermore, all possible network output responses may be obtained by formulating an input stimulus vector of the form $\langle (tt \ldots t)_2|$ allowing for a convenient method to determine the "total network response." Using $\langle t|$ as a component in an input stimulus vector is analogous to the various symbolic simulation methods that are often employed in simulation algorithms where the switching network is modeled using Boolean algebra. Example 4.3 illustrates how multiple output responses can be obtained through the execution of a single vector-matrix multiplication.

Example 4.3 *Network Simulation Examples*
Consider the logic network depicted in Figure 3.5 whose corresponding transfer matrix, **T**, is given in Equation 3.6. If it is desired to obtain all achievable network responses when only the topmost primary input x_1 is constrained to $\langle 0|$, the input stimulus vector is formed as $\langle x_1 x_2| = \langle 0t| = \langle 0| \otimes \langle t| = \begin{bmatrix} 1 & 1 & 0 & 0 \end{bmatrix}$. Evaluating Equation 4.1 with $\langle 0t|$ results in an output response vector of the form:

$$\langle 0t|\mathbf{T} = \begin{bmatrix} 1 & 1 & 0 & 0 \end{bmatrix} \begin{bmatrix} 0 & 1 & 0 & 0 \\ 0 & 0 & 1 & 0 \\ 0 & 0 & 1 & 0 \\ 0 & 0 & 1 & 0 \end{bmatrix} = \begin{bmatrix} 0 & 1 & 1 & 0 \end{bmatrix}$$

$$= \begin{bmatrix} 0 & 1 & 0 & 0 \end{bmatrix} + \begin{bmatrix} 0 & 1 & 0 & 0 \end{bmatrix} = \langle (1)_{10}| + \langle (2)_{10}| = \langle (01)_2| + \langle (10)_2|$$

Thus, with a single calculation, it is determined that the inputs $\langle 00|$ and $\langle 01|$ result in output responses of $\langle 01|$ and $\langle 10|$. Unfortunately, due to commutativity of the vector addition operation, it is not possible to determine which of the two specific input valuations represented by input stimulus $\langle 0t|$ cause one of the specific output valuations to occur. The specific mappings can be obtained through the evaluation of another output response by using input stimulus $\langle 00|$ or $\langle 01|$.

The total output response can be similarly obtained by forming the total input stimulus vector $\langle tt|$ and evaluating Equation 4.1 as:

$$\langle tt|\mathbf{T} = \begin{bmatrix} 1 & 1 & 1 & 1 \end{bmatrix} \begin{bmatrix} 0 & 1 & 0 & 0 \\ 0 & 0 & 1 & 0 \\ 0 & 0 & 1 & 0 \\ 0 & 0 & 1 & 0 \end{bmatrix} = \begin{bmatrix} 0 & 1 & 3 & 0 \end{bmatrix}$$

$$= \begin{bmatrix} 0 & 1 & 0 & 0 \end{bmatrix} + 3 \begin{bmatrix} 0 & 1 & 0 & 0 \end{bmatrix} = \langle (1)_{10}| + 3\langle (2)_{10}| = \langle (01)_2| + 3\langle (10)_2|$$

The total output response is then observed to be a single occurrence of $\langle 01|$ and the occurrence of $\langle 10|$ for the other three responses.   □

The total output response can be very useful since it can be quickly determined which of all possible valuations are not possible for a given logic network.

4.2   TRANSFER MATRIX PROPERTIES

The structure of the transfer matrix representing a switching network can provide useful information concerning characteristics of the network. The transfer matrix can be considered to be a column vector whose components are row vectors that form the set of possible output responses. Likewise, the set of all possible input stimuli are represented as a column vector whose components are each row vectors representing the total collection of unique input stimulus vector valuations. As previously observed, each distinct input stimulus vector valuation is a canonical basis vector $\langle x_i| \in \mathbb{H}^n$. The column vector of all such input stimuli, when written in the usual order appearing in a truth table, form the identity matrix $\mathbf{I}$. Multiplying the transfer matrix $\mathbf{T}$ with a distinct input valuation simply selects the appropriate row vector in $\mathbf{T}$ corresponding to the output response resulting from the specific input stimulus.

A useful result in terms of implementation of these methods is that the transfer matrix is isomorphic to a conventional truth table representation when the network is modeled using switching algebra. This observation allows any of the conventional methods for compact representations of switching functions such as cube lists or binary decision diagrams (BDDs) to also be employed for representation of a switching network transfer matrix. The combination of using efficient transfer matrix representations with the extraction of the transfer matrix from a structural netlist allows the vector space model to be competitive with conventional switching algebraic approaches. The isomorphic relationship between switching algebraic and vector space models is stated in the following observation.

Observation 4.4   *Truth Table Isomorphism*
A truth table representation for a switching network is isomorphic to the corresponding transfer matrix. By definition, we model switching constants in $x_i \in \mathbb{B}$ as row vectors $\langle x_i| \in \mathbb{H}$, that is $0 \to \langle 0|$ and $1 \to \langle 1|$. Thus, a simple substitution of switching values to their corresponding vector space models can be accomplished to derive the transfer matrix. Example 4.5 illustrates this principle for the example switching network given in Figure refnetlist-fig-lab.

$x_1\ x_2$	$f_1\ f_2$	$\langle x_1\|$ $\langle x_2\|$	$\langle f_1\|$ $\langle f_2\|$	$\langle x_1 x_2\|$	$\langle f_1 f_2\|$	$\langle x_1 x_2\|$	$\langle f_1 f_2\|$
0 0	1 0	$\langle 0\|$ $\langle 0\|$	$\langle 1\|$ $\langle 0\|$	$\langle 00\|$	$\langle 10\|$	[1 0 0 0]	[0 0 1 0]
0 1	0 1	$\langle 0\|$ $\langle 1\|$	$\langle 0\|$ $\langle 1\|$	$\langle 01\|$	$\langle 01\|$	[0 1 0 0]	[0 1 0 0]
1 0	0 1	$\langle 1\|$ $\langle 0\|$	$\langle 0\|$ $\langle 1\|$	$\langle 10\|$	$\langle 01\|$	[0 0 1 0]	[0 1 0 0]
1 1	0 1	$\langle 1\|$ $\langle 1\|$	$\langle 0\|$ $\langle 1\|$	$\langle 11\|$	$\langle 01\|$	[0 0 0 1]	[0 1 0 0]

$$\mathbf{T} = \begin{bmatrix} 0 & 0 & 1 & 0 \\ 0 & 1 & 0 & 0 \\ 0 & 1 & 0 & 0 \\ 0 & 1 & 0 & 0 \end{bmatrix}$$

Figure 4.1: Truth table isomorphism example.

Example 4.5   *Truth Table Isomorphism Example*
Figure 4.1 depicts the truth table for the network in Figure 3.5 on the left and the corresponding transfer matrix on the right through a simple substitution of switching values in $\mathbb{B}$ to their corresponding vector model values in $\mathbb{H}$. Individual valuations in $\mathbb{H}$ are combined into a single higher-dimensioned vector in $\mathbb{H}^m$ through the application of the outer product to expand the dimensionality of the vector space.

□

Example 4.3 indicates that it is not possible for the logic network in Figure 3.5 to ever provide output responses in the form of $\langle 00|$ or $\langle 11|$. Although this conclusion is based upon the results of network output response calculations, it is possible to determine the number of and to identify the non-possible output responses directly from the form of the transfer matrix.

Lemma 4.6   Allowable Output Response of a Logic Network
The number of possible distinct output responses of a logic network characterized by a transfer matrix $\mathbf{T}$ *is equivalent to* $2^m - N_{null}$ *where* N_{null} *is the number of null column vectors comprising* $\mathbf{T}$.

Proof. The non-standard approach of indexing column and row vectors of a matrix by using indices in the range $[0, 2^n - 1]$ is employed in this work since it allows the row vector index value of a transfer matrix to directly yield the corresponding input stimulus vector that would cause a particular output response to occur. From the preceding discussion, each distinct input stimulus vector is a canonical basis vector $\langle i| \in \mathbb{H}^n$. Thus, the output response calculation simply selects a row vector in $\mathbf{T}$ as the corresponding output response vector. If a column vector of $\mathbf{T}$ is null, it indicates that no possible input stimulus will result in that particular output response denoted by $\langle j|$ where j is the column vector index of $\mathbf{T}$ where $j = [0, 2^m - 1]$ for a network with m primary outputs. The total number of null column vectors comprising $\mathbf{T}$ is denoted as N_{null}, therefore the total number of permissible output responses of a switching network characterized by $\mathbf{T}$ is $2^m - N_{null}$.

□

A single-output switching network is tautological when the output response is $\langle 1|$ regardless of the primary input assignment. Due to the observations of the structure of a transfer matrix, it is apparent that a tautological network must have a characterizing transfer matrix whose column vector with index $j = 0$ is $\langle \varnothing \varnothing \ldots \varnothing|$ and whose $j = 1$ indexed column vector is $\langle tt \ldots t|$. Likewise, a contradictory switching network is the inverse of a tautological network and always yields an output response of $\langle 0|$. The contradictory network must have a characterizing transfer matrix whose $j = 0$ column vector is the total vector $\langle tt \ldots t|$ and whose $j = 1$ vector is null.

In general, the form of a transfer matrix is $\mathbf{T} = [t_{ij}]_{N \times M}$ where $log_2(N) = n$ the number of primary inputs and $log_2(M) = m$, the number of primary outputs. In the case where $N = M$, the network is comprised of an equal number of primary inputs and outputs. Furthermore, when $\mathbf{T}$ is of full rank and $N = M$, a bijective mapping is present since the collection of input stimuli are each uniquely mapped to a corresponding output response. This special class of switching networks are said to be *logically reversible* and is a subject of interest in the research community due to the results of Landauer [18] which state that such networks when used to process information do not dissipate power due to information loss.

Depending upon the implementation technology, a reversible switching network may also allow an output response to be applied to the physical implementation of a network and the primary inputs to then produce the resulting input stimulus. Such is not the case for common electronic switching networks implemented in a technology such as static CMOS electronic transistor networks, thus this class of networks is logically reversible, but not physically reversible. Quantum logic networks are necessarily both logically and physically reversible due to the laws of quantum mechanics. For the case of quantum logic networks, other properties of their transfer matrices are also in place such as the matrices being unitary and being comprised of complex-valued coefficients. The vector space model presented in this work for conventional switching networks can thus be viewed as a superset of the special case of reversible switching networks and provides a convenient mathematically unifying theory for modeling conventional reversible and quantum logic networks with the general case of irreversible switching networks where $N \neq M$ is commonly encountered. This unification of mathematical modeling is one advantage of the approach described here as it provides a means for comparing network functionality among these different forms of physical implementation.

4.3   THE PSEUDO-INVERSE OF A TRANSFER MATRIX AND NETWORK JUSTIFICATION

Several EDA tasks require the determination of an input stimulus vector given a characterization of the network and an output response, referred to as the *justification* problem. Logic network justification is useful in multiple design and analysis applications, including synthesis, verification, and test. In terms of automatic test pattern generation (ATPG) algorithms including the D-algorithm [32], PODEM [33], and FAN [34], justification is a core technique. It is also inherently related to the satisfiability problem (SAT). Furthermore, reverse logic implications are a special case where *all possible* solutions require that a particular logic value (or set of logic values) be assigned to a particular input (or set of inputs).

Justification during ATPG is historically performed by successively assigning logic values to upstream gates or primary inputs and then propagating the results downstream through forward implication until the desired downstream assignment has been justified. However, reconvergent fanout may produce conflicts among selected value assignments. This requires that previously assigned nodes be assigned a different value and thus requires *backtracking*. Similarly, a widely applied technique for finding implications involves the construction of a binary learning tree whose traversal, in conjunction with a recursive learning algorithm, allows assignment of logic values at various logic network nodes [35]. In the learning approach, a backtracking algorithm is also used.

Performing justification using the vector space model is accomplished without the need for successive logic assignments, backtracking, recursive learning, or other iterative methods since a network is justified through a single vector-matrix multiplication operation. Network justification using the vector space model requires the solution of $\langle f_i| = \langle x_i|\mathbf{T}$ for $\langle x_i|$ when $\mathbf{T}$ and $\langle f_i|$ are known. The justification problem as described within the framework of the vector space model is simply the determination of the inverse $\mathbf{T}^{-1}$ if the network is reversible.

By definition, the inverse $\mathbf{T}^{-1}$ always exists and is unique for reversible networks since $\mathbf{T}$ is square and of full rank. Unfortunately, a large number of switching networks currently in use are irreversible and thus the justification problem cannot be solved in this manner due to the non-existence of $\mathbf{T}^{-1}$. Mathematically, the solution of $\langle f_i| = \langle x_i|\mathbf{T}$ for $\langle x_i|$ when $N \neq M$ results in two cases. When $N > M$, the system is over-constrained and multiple solutions exist. Conversely when $N < M$, the system is under-constrained. One approach for determination of $\langle x_i|$ when $N \neq M$ is to utilize the Moore-Penrose pseudo inverse of $\mathbf{T}$ denoted as $\mathbf{T}^+$ [6]. Equation 4.2 gives the form of the pseudo-inverse.

$$\mathbf{T}^+ = \begin{cases} (\mathbf{T}^*\mathbf{T})^{-1}\mathbf{T}^*, & N > M, \text{overspecified} \\ \mathbf{T}^*(\mathbf{T}\mathbf{T}^*)^{-1}, & N < M, \text{underspecified} \end{cases} \tag{4.2}$$

In Equation 4.2, $\mathbf{T}^*$ denotes the complex-transpose or Hermitian of $\mathbf{T}$. For the binary switching networks considered here, the components of $\mathbf{T}$ are always real-valued hence $\mathbf{T}^* = \mathbf{T}^T$ where $\mathbf{T}^T$ denotes the transpose. The use of the pseudo-inverse allows for a solution to the justification problem in the form $\langle x_i| = \langle f_i|\mathbf{T}^+$. In the case of over constrained systems, multiple solutions exist and the solution obtained using the pseudo-inverse is the single solution that has minimal relative error to all the solutions in a least-squared sense. Fortunately, the error is by definition zero-valued since it is known that the solutions are all of the form of $\langle 0|$ or $\langle 1|$, hence the justification solution using $\mathbf{T}^T$ is the exact solution. Likewise, in the case of under constrained systems, the pseudo-inverse provides a best-fit solution in the least-squared error sense to the exact solution. Again, because the exact solution is known to be either $\langle 1|$ or $\langle 0|$, the least-squared error is zero-valued. This fortunate set of circumstances allows the pseudo-inverse to be employed resulting in an exact solution to the justification problem. Equation 4.2 also contains a term in the form of $\mathbf{T}^T\mathbf{T}$ that is the square *Gram matrix*, or *Gramian* of $\mathbf{T}$.

Definition 4.7   *Gram Matrix*

The Gram matrix of $\mathbf{T}$ is denoted as $gram(\mathbf{T}) = \mathbf{T}^T\mathbf{T}$ and is always invertible. $gram(\mathbf{T})$ is a positive semi-definite Hermitian matrix since $\langle x|\mathbf{T}^T\mathbf{T}|x\rangle \geq 0$ for all non-null $|x\rangle$. The Gramian has several useful properties. For the real transfer matrices considered in this book, the corresponding Gramians are symmetric, $[gram(\mathbf{T})]^T = gram(\mathbf{T})$. Other properties include the fact that the rank of $gram(\mathbf{T})$ is equivalent to the rank of $\mathbf{T}$ and, when $\mathbf{T}$ is square, the determinant of $gram(\mathbf{T})$ is equivalent to the determinant of $\mathbf{T}$. Eigenvectors $|v_i\rangle$ exist for $gram(\mathbf{T})$, and the corresponding vector $\mathbf{T}|v_i\rangle$ is an eigenvector of $\mathbf{T}\mathbf{T}^T$. Each of these two eigenvectors has the same eigenvalue whose positive square root is a *singular value* of $\mathbf{T}$.   □

For the class of possible matrices representing a switching network $\mathbf{T}$, the Gramian is always in the form of a diagonal matrix, and the diagonal elements yield further information about the switching network as given in following Lemmas and Theorems.

Lemma 4.8   Gramian of Single-output Network

The diagonal elements of the 2×2 Gramian of a matrix $\mathbf{T}$ characterizing an n-input, single output switching network are the number of maxterms and minterms of the network.

Proof. In general, the form of the transfer matrix $\mathbf{T}$ for an n-input, single-output switching network is of the form of a $2^n \times 2$ matrix. This can be written as a 2-dimensional row vector whose components are the column vectors $|f_0\rangle$ and $|f_1\rangle$. Due to the definition of $\langle 0|$ and $\langle 1|$, the sum of the components of column vector $|f_0\rangle$ yields the number of maxterms, N_{max}, and the sum of components of column vector $|f_1\rangle$ yields the number of minters, N_{min}. Furthermore, the inner product of each column vector $|f_i\rangle$ with itself, $\langle f_i|f_i\rangle$, is equivalent to the sum of the components. The transfer matrix $\mathbf{T}$ has the form

$$\mathbf{T} = \left[\ |f_0\rangle\ \ |f_1\rangle\ \right] = \begin{bmatrix} f_{0,0} & f_{1,0} \\ f_{0,1} & f_{1,1} \\ f_{0,2} & f_{1,2} \\ \cdots & \cdots \\ f_{0,2^n-1} & f_{1,2^n-1} \end{bmatrix}$$

where the ith row vector is of the form $\langle f_i| = \left[\ f_{0,i}\ \ f_{1,i}\ \right]$ and is either a maxterm, $\langle 0|$, or a minterm, $\langle 1|$. Furthermore, since $(f_{0,i}, f_{1,i}) \in \mathbb{B}$ and $f_{0,i} + f_{1,i} = 1$, $\mathbf{T}$ can be rewritten as

$$\mathbf{T} = \begin{bmatrix} |1-f_1\rangle & |f_1\rangle \end{bmatrix} = \begin{bmatrix} 1-f_{1,0} & f_{1,0} \\ 1-f_{1,1} & f_{1,1} \\ 1-f_{1,2} & f_{1,2} \\ \cdots & \cdots \\ 1-f_{1,2^n-1} & f_{1,2^n-1} \end{bmatrix}.$$

The Gramian of $\mathbf{T}$ becomes

$$\mathbf{T}^T\mathbf{T} = \begin{bmatrix} 1-f_{1,0} & 1-f_{1,1} & 1-f_{1,2} & \cdots & 1-f_{1,2^n-1} \\ f_{1,0} & f_{1,1} & f_{1,2} & \cdots & f_{1,2^n-1} \end{bmatrix} \begin{bmatrix} 1-f_{1,0} & f_{1,0} \\ 1-f_{1,1} & f_{1,1} \\ 1-f_{1,2} & f_{1,2} \\ \cdots & \cdots \\ 1-f_{1,2^n-1} & f_{1,2^n-1} \end{bmatrix}$$

$$= \begin{bmatrix} 2^n - \langle f_1|f_1\rangle & 0 \\ 0 & \langle f_1|f_1\rangle \end{bmatrix} = \begin{bmatrix} 2^n - N_{min} & 0 \\ 0 & N_{min} \end{bmatrix} = \begin{bmatrix} N_{max} & 0 \\ 0 & N_{min} \end{bmatrix}.$$

$\square$

Theorem 4.9   Self-inner Product of Column Vector of $\mathbf{T}$ ($\|L_2(|f_i\rangle)\|^2$)
The inner product of the ith column vector of $\mathbf{T}$ with itself yields the number of output response valuations equivalent to $\langle i|$, or $\|L_2(|f_i\rangle)\|^2 = N_i$.

Proof. A network with m primary outputs may be described with m truth tables or as a single truth table with m output columns as shown in Figure 4.1. When the m output columns are expressed as a single column of 2^m row vectors, each row vector is of the form of a canonical basis vector, $\langle i|$ and contains a single unity-valued component in the ith position with all other components being zero-valued. Using this observation and generalizing the derivation in the proof of Lemma 4.8 yields the result of the theorem. $\square$

Example 4.10 illustrates how the pseudo-inverse of a transfer matrix can be used to justify the inputs of the switching network in Figure 3.5.

Example 4.10   *Justification Example with Pseudo-inverse of $\mathbf{T}$*
Using Equation 4.2, the pseudo-inverse is calculated as

$$\mathbf{T}^+ = \begin{bmatrix} 0 & 0 & 0 & 0 \\ 0 & \frac{1}{3} & 0 & 0 \\ 0 & 0 & 1 & 0 \\ 0 & 0 & 0 & 0 \end{bmatrix} \begin{bmatrix} 0 & 0 & 0 & 0 \\ 0 & 1 & 1 & 1 \\ 1 & 0 & 0 & 0 \\ 0 & 0 & 0 & 0 \end{bmatrix} = \begin{bmatrix} 0 & 0 & 0 & 0 \\ 0 & \frac{1}{3} & \frac{1}{3} & \frac{1}{3} \\ 1 & 0 & 0 & 0 \\ 0 & 0 & 0 & 0 \end{bmatrix}$$

Assuming the output response is $\langle f_1 f_2| = \langle 10|$, the corresponding input stimulus can be computed using the pseudo-inverse to justify $\langle 10|$ as

$$\langle x_1 x_2| = \langle 10|\mathbf{T}^+ = \begin{bmatrix} 0 & 0 & 1 & 0 \end{bmatrix} \begin{bmatrix} 0 & 0 & 0 & 0 \\ 0 & \frac{1}{3} & \frac{1}{3} & \frac{1}{3} \\ 1 & 0 & 0 & 0 \\ 0 & 0 & 0 & 0 \end{bmatrix} = \begin{bmatrix} 1 & 0 & 0 & 0 \end{bmatrix} = \langle 00|.$$

Accordingly, the justification of $\langle f_1 f_2| = \langle 01|$ is calculated as

$$\langle x_1 x_2| = \langle 01|\mathbf{T}^+ = \begin{bmatrix} 0 & 1 & 0 & 0 \end{bmatrix} \begin{bmatrix} 0 & 0 & 0 & 0 \\ 0 & \frac{1}{3} & \frac{1}{3} & \frac{1}{3} \\ 1 & 0 & 0 & 0 \\ 0 & 0 & 0 & 0 \end{bmatrix} = \begin{bmatrix} 0 & \frac{1}{3} & \frac{1}{3} & \frac{1}{3} \end{bmatrix}$$

$$= \frac{1}{3}(\langle 01| + \langle 10| + \langle 11|).$$

This result indicates that the three different input stimulus vectors $\langle 01| + \langle 10| + \langle 11|$ justify $\langle f_1 f_2| = \langle 01|$. The coefficient $\frac{1}{3}$ is a scaling side-effect of the pseudo-inverse required to ensure that $\mathbf{T}\mathbf{T}^+ = \mathbf{T}^+\mathbf{T} = \mathbf{I}$.

When an infeasible output response is justified, the null vector results. Consider the output assignment of $\langle f_1 f_2| = \langle 11|$. The justification calculation becomes

$$\langle x_1 x_2| = \langle 11|\mathbf{T}^+ = \begin{bmatrix} 0 & 0 & 0 & 1 \end{bmatrix} \begin{bmatrix} 0 & 0 & 0 & 0 \\ 0 & \frac{1}{3} & \frac{1}{3} & \frac{1}{3} \\ 1 & 0 & 0 & 0 \\ 0 & 0 & 0 & 0 \end{bmatrix} = \begin{bmatrix} 0 & 0 & 0 & 0 \end{bmatrix} = \langle \varnothing\varnothing|.$$

Finally, use of the total vector can conveniently be used to justify subsets of output responses. As an example, if it is desired to justify the case $\langle f_2| = \langle 1|$ regardless of the valuation of $\langle f_1|$, an output response vector of the form $\langle t1|$ can be used in the justification calculation. Likewise, the total justification can be computed using $\langle f_1 f_2| = \langle tt|$

$$\langle x_1 x_2| = \langle tt|\mathbf{T}^+ = \begin{bmatrix} 1 & 1 & 1 & 1 \end{bmatrix} \begin{bmatrix} 0 & 0 & 0 & 0 \\ 0 & \frac{1}{3} & \frac{1}{3} & \frac{1}{3} \\ 1 & 0 & 0 & 0 \\ 0 & 0 & 0 & 0 \end{bmatrix} = \begin{bmatrix} 1 & \frac{1}{3} & \frac{1}{3} & \frac{1}{3} \end{bmatrix}$$

$$= \langle 00| + \frac{1}{3}(\langle 01| + \langle 10| + \langle 11|).$$

In this latter case, the scaling coefficients of the input stimuli vectors yield useful information regarding the distribution of distinct input stimuli. Because $\langle 01|$, $\langle 10|$, and $\langle 11|$ have a common multiple of $\frac{1}{3}$, it can be concluded that they correspond to the same output response vector. □

4.4 THE JUSTIFICATION MATRIX

The calculation of the pseudo-inverse is easily accomplished due to the fact that the Gramian of $\mathbf{T}$ is diagonal. Therefore, the inverse $[gram(\mathbf{T})]^{-1} = (\mathbf{T}^T \mathbf{T})^{-1}$ is also diagonal where the diagonal components of the inverse are simply the multiplicative inverses of the diagonal values of $gram(\mathbf{T})$. Nevertheless, in spite of the ease of obtaining $[gram(\mathbf{T})]^{-1}$, the computation of the pseudo-inverse can be avoided altogether by using the transpose of $\mathbf{T}$ in place of $\mathbf{T}^+$. This observation leads to the definition of the *justification matrix* $\mathbf{T}^J$.

Definition 4.11 The "justification Matrix" for a switching network characterized by a transfer function $\mathbf{T}$ is denoted as $\mathbf{T}^J$ and is defined to be the transpose of the transfer matrix.

$$\mathbf{T}^J = \mathbf{T}^T \tag{4.3}$$

The advantage of using the justification matrix is that a justification computation may be accomplished without computing a pseudo-inverse matrix. The disadvantage is that scaling coefficients are no longer present in the results, thus it is possible to determine a set of justifying input values for a given output response, but it is no longer possible to determine how they are distributed among all possible input stimuli. Example 4.12 illustrates the use of the justification matrix, $\mathbf{T}^J$.

Example 4.12 *Justification Example with Pseudo-inverse of* $\mathbf{T}$
Using Equation 4.3, the switching network depicted in Figure 3.5 can be justified for the output response constraint of $\langle f_2| = \langle 1|$ as

$$\langle x_1 x_2| = \langle t1|\mathbf{T}^J = \begin{bmatrix} 0 & 1 & 0 & 1 \end{bmatrix} \begin{bmatrix} 0 & 0 & 0 & 0 \\ 0 & 1 & 1 & 1 \\ 1 & 0 & 0 & 0 \\ 0 & 0 & 0 & 0 \end{bmatrix} = \begin{bmatrix} 0 & 1 & 1 & 1 \end{bmatrix}$$

$$= \langle 01| + \langle 10| + \langle 11|.$$

The result of this justification calculation indicates that the three resulting input stimuli will cause either $\langle 01|$ or $\langle 11|$ to result since $\langle t1|$ covers both of these output responses. In reality, $\langle 11|$ is an infeasible output response and the three resulting justified input stimuli all result in only one of two specified output responses, $\langle f_1 f_2| = \langle 01|$. This result is easily verified with a simulation computation. □

CHAPTER 5

MVL Switching Networks

5.1 MV INFORMATION REPRESENTATION IN THE VECTOR SPACE

The previous chapters assume that the switching networks operate over two-valued or binary information. While this is certainly the most prevalent form of switching circuitry in use, there is a considerable amount of research and interest in non-binary systems in which $r > 2$. The general term for non-binary systems with a finite number of states is that of *Multiple Valued Logic* (MVL). We shall refer to switching circuits that utilize $r > 2$ states as *Multiple Valued Switching Networks* (MVSN). The vector space model for information representation is directly extendable for MVSN. Table 5.1 contains the canonical basis vectors for radices $r = \{2, 3, 4, 5\}$ to illustrate how the vector model can be extended.

Table 5.1: Canonical basis vectors for $r = \{2, 3, 4, 5\}$

RADIX r	VALUE $\langle 0|$	VALUE $\langle 1|$	VALUE $\langle 2|$	VALUE $\langle 3|$	VALUE $\langle 4|$
2	[1 0]	[0 1]	n/a	n/a	n/a
3	[1 0 0]	[0 1 0]	[0 0 1]	n/a	n/a
4	[1 0 0 0]	[0 1 0 0]	[0 0 1 0]	[0 0 0 1]	n/a
5	[1 0 0 0 0]	[0 1 0 0 0]	[0 0 1 0 0]	[0 0 0 1 0]	[0 0 0 0 1]

This generalization of the vector space model is advantageous as compared to switching algebraic formulations. As described in [26], Boolean algebras are not functionally complete when the radix r is not a power of two. Thus, switching models must utilize alternative algebraic formulations such as that of Emil Post [30]. The application of the vector model as an alternative to switching models for ternary $r = 3$ MVSNs was first described in [27] and applications based on the technique appeared in [28] and [29].

In addition to the canonical basis vectors that model specific valuations of network data given in Table 5.1, the concept of the null vector $\langle \varnothing|$ is also useful regardless of the radix r. When $r > 2$, the concept of the *total vector* $\langle t|$ can be generalized and several intermediate values can be formulated that partially cover subsets of values. This leads to several intermediate values that are of use in various modeling applications and as r increases, the number of covering values likewise

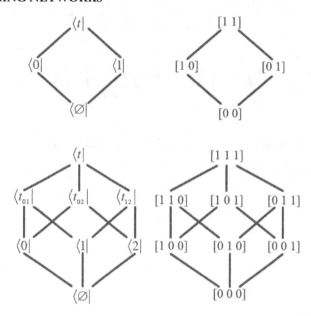

Figure 5.1: Hasse diagrams for $r = 2$ and $r = 3$.

increases. In terms of a ternary $r = 3$ system there are three distinct vector constants $\langle 0|$, $\langle 1|$, and $\langle 2|$; however, there are also five other constants that are defined and used, denoted as $\langle t|$, $\langle t_{01}|$, $\langle t_{02}|$, $\langle t_{12}|$, and $\langle \varnothing|$. $\langle t|$ represents the simultaneous valuations of $\langle 0|$, $\langle 1|$, and $\langle 2|$ and is given by $\langle t| = \langle 0| + \langle 1| + \langle 2|$ where the + operator denotes vector space addition. Likewise, the constants $\langle t_{12}|$, $\langle t_{02}|$, and $\langle t_{01}|$ represent the simultaneous valuations of two logic values $\langle t_{01}| = \langle 0| + \langle 1|$, $\langle t_{02}| = \langle 0| + \langle 2|$, and $\langle t_{12}| = \langle 1| + \langle 2|$. In contrast, $\langle \varnothing|$ represents the absence of all logic values and is given as $\langle \varnothing| = \begin{bmatrix} 0 & 0 & 0 \end{bmatrix}$. $\langle \varnothing|$ should not be confused with the logic-0 value, $\langle 0| = \begin{bmatrix} 1 & 0 & 0 \end{bmatrix}$. The collection of vector constants can be expressed as a Hasse diagram as shown in Figure 5.1 [26]. Two versions of the Hasse diagram are depicted for both $r = 2$ and $r = 3$ using bra- and the row-vector notation. The five additional constants can arise during calculations using transfer function representations of logic networks.

The Hasse diagrams provide a graphical representation of covering values and can be used to graphically depict addition and subtraction operations over the set of switching network values when the additional values of $\langle \varnothing|$ and $\langle t_i|$ are included. As an example, $\langle 0| + \langle 1| + \langle 2| = \langle t|$, $\langle 0| + \langle 2| = \langle t_{02}|$, $\langle t| - \langle t_{01}| = \langle 2|$, $\langle t_{12}| - \langle t_{12}| = \langle \varnothing|$, $\langle 1| - \langle 1| = \langle \varnothing|$, and $\langle 0| = \langle \varnothing| = \langle 0|$.

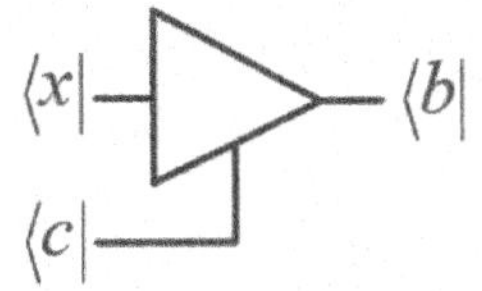

x	c	b
0	0	Z
0	1	0
1	0	Z
1	1	1

$\langle x \vert$	$\langle c \vert$	$\langle b \vert$
$\langle 0 \vert$	$\langle 0 \vert$	$\langle 2 \vert$
$\langle 0 \vert$	$\langle 1 \vert$	$\langle 0 \vert$
$\langle 1 \vert$	$\langle 0 \vert$	$\langle 2 \vert$
$\langle 1 \vert$	$\langle 1 \vert$	$\langle 1 \vert$

$$\mathbf{B} = \begin{bmatrix} 0 & 0 & 1 \\ 1 & 0 & 0 \\ 0 & 0 & 1 \\ 0 & 1 & 0 \end{bmatrix}$$

Figure 5.2: Tri-state buffer symbol, truth tables, and transfer matrix.

5.2 EXTENSIONS TO BINARY SWITCHING NETWORKS

During the development of EDA tools and methods, particularly HDLs, it became apparent that more than two values are needed for many tasks such as simulation and synthesis. This need led to standardizations specific to the HDLs Verilog and VHDL [16] and [17]. Many other switching network descriptions use alternative values such as the – symbol found in .pla files. The vector space model extension for $r > 2$ overcomes the need for these customized extensions since the postulates and axioms of linear algebra are sufficient to represent vector space information models regardless of the radix value r.

A very common example of a network element requiring more than two values is the binary network element commonly referred to as the *tri-state buffer*. Tri-state buffers are used for a variety of multiplexing and bus access systems. In addition to switching values $\langle 0 \vert$ and $\langle 1 \vert$, the tri-state buffer is capable of producing a high-impedance output that can be modeled as $\langle 2 \vert$. Figure 5.2 contains the graphical depiction of a tri-state buffer, the switching algebra truth table, the vector space model truth table, and, by truth table isomorphism, the transfer matrix.

Tri-state buffers may be constructed with a variety of different polarities on both the input and output ports enabling the use of both active-high and active-low electrical signals. Four examples of these different variations of tri-state buffers are given in Figure 5.3 along with their respective transfer matrices.

One example of the use of the tri-state buffer is in the configuration of a two-to-one multiplexer configuration. This configuration is also of interest with respect to the vector modeling approach due to the fact that it includes a fanin structure. Figure 5.4 depicts the circuit diagram for a tri-state multiplexer with partition cuts shown.

The transfer function can be written in factored form as

$$\mathbf{T}_{mux} = (\mathbf{I} \otimes \mathbf{F} \mathbf{O} \otimes \mathbf{I})(\mathbf{B}_{nc} \otimes \mathbf{B})(\mathbf{F} \mathbf{I})$$

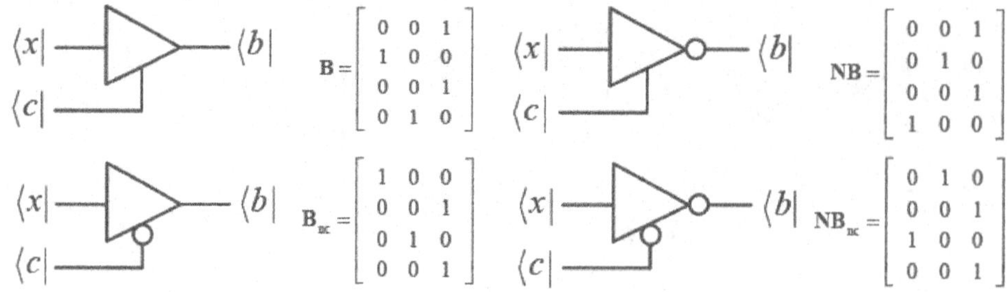

$\langle x| \!-\!\triangleright\!-\! \langle b| \quad \mathbf{B} = \begin{bmatrix} 0 & 0 & 1 \\ 1 & 0 & 0 \\ 0 & 0 & 1 \\ 0 & 1 & 0 \end{bmatrix} \qquad \langle x| \!-\!\triangleright\!\circ\!-\! \langle b| \quad \mathbf{NB} = \begin{bmatrix} 0 & 0 & 1 \\ 0 & 1 & 0 \\ 0 & 0 & 1 \\ 1 & 0 & 0 \end{bmatrix}$
$\langle c|$ $\langle c|$

$\langle x| \!-\!\triangleright\!-\! \langle b| \quad \mathbf{B}_{\mathrm{\propto}} = \begin{bmatrix} 1 & 0 & 0 \\ 0 & 0 & 1 \\ 0 & 1 & 0 \\ 0 & 0 & 1 \end{bmatrix} \qquad \langle x| \!-\!\triangleright\!\circ\!-\! \langle b| \quad \mathbf{NB}_{\mathrm{\propto}} = \begin{bmatrix} 0 & 1 & 0 \\ 0 & 0 & 1 \\ 1 & 0 & 0 \\ 0 & 0 & 1 \end{bmatrix}$
$\langle c|$ $\langle c|$

Figure 5.3: Tri-state buffers with various input/output polarities.

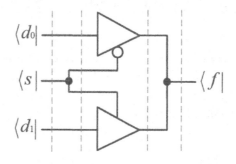

Figure 5.4: Two-to-one multiplexer with tri-state buffer.

Although it is impossible for the multiplexer to produce a high-impedance output, all transfer matrices must be modeled within an $r = 3$ system. To illustrate this concept, the explicit form of the transfer matrix for the tri-state buffer-based multiplexer is

$$\mathbf{T}_{mux} = \left(\begin{bmatrix} 1 & 0 & 0 \\ 0 & 1 & 0 \\ 0 & 0 & 1 \end{bmatrix} \otimes \begin{bmatrix} 1 & 0 & 0 & 0 & 0 & 0 & 0 & 0 & 0 \\ 0 & 0 & 0 & 0 & 1 & 0 & 0 & 0 & 0 \\ 0 & 0 & 0 & 0 & 0 & 0 & 0 & 0 & 1 \end{bmatrix} \otimes \begin{bmatrix} 1 & 0 & 0 \\ 0 & 1 & 0 \\ 0 & 0 & 1 \end{bmatrix} \right)$$

$$\left(\begin{bmatrix} 1 & 0 & 0 \\ 0 & 0 & 1 \\ 0 & 1 & 0 \\ 0 & 0 & 1 \end{bmatrix} \otimes \begin{bmatrix} 0 & 0 & 1 \\ 1 & 0 & 0 \\ 0 & 0 & 1 \\ 0 & 1 & 0 \end{bmatrix} \right) \left(\begin{bmatrix} 1 & 0 & 0 \\ 0 & 0 & 0 \\ 0 & 0 & 0 \\ 0 & 0 & 0 \\ 0 & 1 & 0 \\ 0 & 0 & 0 \\ 0 & 0 & 0 \\ 0 & 0 & 0 \\ 0 & 0 & 1 \end{bmatrix} \right)$$

$$= \begin{bmatrix} 1 & 0 & 0 \\ 1 & 0 & 0 \\ 1 & 0 & 0 \\ 0 & 1 & 0 \\ 0 & 1 & 0 \\ 0 & 1 & 0 \\ 1 & 0 & 0 \\ 0 & 1 & 0 \end{bmatrix}.$$

The rightmost column vector of $\mathbf{T}_{mux}$ is equivalent to $|\varnothing\rangle$ since it is impossible for the two-to-one multiplexer to produce an output response value of Z as long as all primary inputs have values from the set $\{\langle 0|, \langle 1|\}$ when the physical implementation of the switching circuit is accomplished with conventional static-CMOS electronic circuitry. Because the switching network contains three-state elements, all intermediate transfer matrices must be formulated to account for this third state high-impedance state of Z. All primary input values must also be formulated as a ternary value regardless of the fact that the circuit is practically intended to only model the case when $\{\langle d_0|, \langle s|, \langle d_i|\}$ have values of either $\langle 0| = \begin{bmatrix} 1 & 0 & 0 \end{bmatrix}$ or $\langle 1| = \begin{bmatrix} 0 & 1 & 0 \end{bmatrix}$. This is important to note since the calculation of an output response using the transfer matrix $\mathbf{T}_{mux}$ will mathematically support the case of one or more of the inputs at high-impedance Z, but the corresponding output response will likely not accurately model a physical implementation. When primary inputs are set to the high-impedance state, it is implied that the underlying physical technology for constructing the switching network is electronic and the corresponding output response is dependent upon the type and configuration of the electronic components.

5.3   GENERAL TERNARY SWITCHING NETWORKS

Ternary switching functions are traditionally modeled with an algebraic structure consisting of a discrete set of switching constants $\{0, 1, 2\}$ and an appropriate set of operators. This is the more general case where $\langle 2 |$ is considered as a third state value rather than interpreting it as high-impedance Z as is the case for the two-to-one multiplexer in Figure 5.4. General ternary switching networks may represent implementations using either electronic or some other technology with a corresponding input/output characteristic modeled by ternary switching functions that have three distinct output state values. A general ternary switching network is then considered as an interconnection of symbols or gates whose functionality is modeled by the operators within the algebraic structure. A common set of ternary switching network elements are the *MIN*, *MAX*, and *Literal Selection* gates whose corresponding functionalities provide for the specification of a functionally complete algebra with constants. For conciseness we refer to the *Literal Selection* gate as a J_i gate where i denotes the polarity of the selected literal, $i \in \{0, 1, 2\}$ as defined in [26].

Within the framework of the linear algebraic model, we use three-dimensional vectors to represent ternary state values instead of the integers $\{0, 1, 2\}$ and we model a ternary switching function or a ternary switching network by a characterizing transfer matrix. The transfer matrix transforms the switching network input vectors to corresponding output vectors. This linear algebraic model of a ternary switching network or function is an alternative to the more commonly employed models.

Ternary Network Elements

Any of a variety of two- and one-place operators can be defined and represented by network gates within a ternary switching network. For the algebraic construction to enable representation of all possible functions, some primitive set of operators is required such that all other possible operators can be expressed as a function of the primitives and constants [26]. In this work, we utilize the ternary MIN, MAX, and J_i operators as logic primitives. The MIN gate is a generalization of the binary AND gate and produces an output response that is the arithmetic minimum of the set of all input values. Likewise, the MAX gate is a generalization of the binary inclusive-OR gate producing the maximum-value among all input stimulus values. Because MIN and MAX are generalizations of the $r - 2$ AND and OR gates, the same switching gate symbol is used to denote them in circuit diagram views of MVSNs.

For the MVSN to have the capability to represent any arbitrarily specified function, it is also required that a set of unary operations or gates be present within the MVSN. Unlike the binary case where there are only $r^r = 2^2 = 4$ possible unary operators, there are $r^r = 3^3 = 27$ possible unary operations and the inclusion of various subsets of these possible operators along with the binary MIN and MAX functions has resulted in a variety of different collections of operators, or, from a switching theory point of view, algebras.

A very common set of operators are the *literal selection* gates that are usually included with the MIN and MAX gates to form an algebra known as the "Post algebra." These three gates are

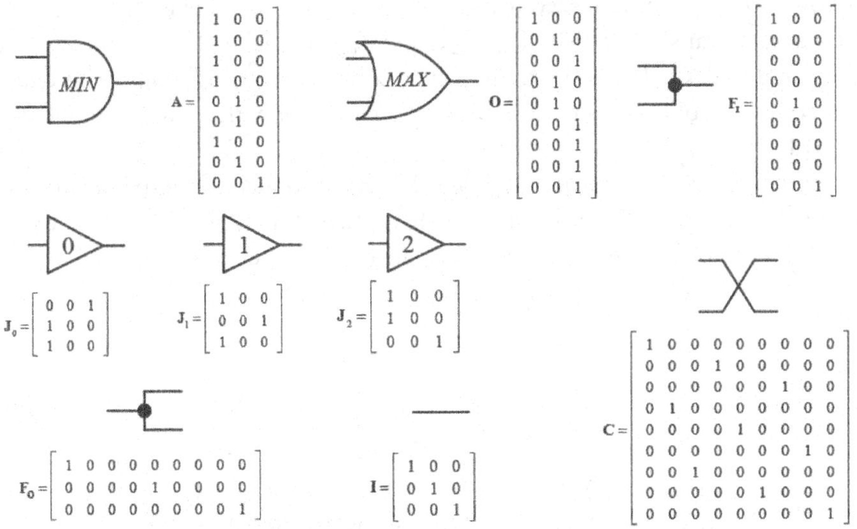

Figure 5.5: Set of functionally complete ternary MVSN gates and circuit structures.

denoted as J_i where $i = \{0, 1, 2\}$ specifies which of the three specific literal gates are employed. They are denoted as switching circuit symbols using the triangular buffer symbol annotated with one of the integers $\{0, 1, 2\}$ and produce an output of 2, the maximum-valued switching constant when the input stimulus is the same as i. Otherwise, the literal selection gate produces an output of 0 if the input stimulus is not equivalent to i. Figure 5.3 contains the symbols for each of these gates with their corresponding transfer matrices.

Transfer Matrices for Ternary MVSNs

The derivations for binary $r = 2$ switching network transfer matrices are easily generalized for the case of ternary $r = 3$ and higher radices. Details of these derivations are given in [27]. The form of the transfer matrix as a sum of vector projection matrices is given in Equation 5.1.

$$\mathbf{T} = \sum_{i=0}^{3^n-1} |x_i\rangle\langle f_i| \tag{5.1}$$

Furthermore, the property of truth isomorphism holds as does the methodology for extraction of the transfer matrix from a structural representation of a ternary MVSN. As an example of the application of the extraction of a transfer matrix from a structural representation, we consider the example ternary MVSN depicted in Figure 5.3 shown with partition cuts computed via levelization as dashed lines, a Verilog-like structural description, and a graphical depiction of the distributed transfer matrices. The logic network is partitioned into cascades of parallel components as shown by the vertical dashed lines. Each serial cascade stage is labeled as ϕ_1, ϕ_2, and ϕ_3 with corresponding transfer matrices $\mathbf{T}_{\phi_1}$, $\mathbf{T}_{\phi_2}$, and $\mathbf{T}_{\phi_3}$. $\mathbf{T}_{\phi_1}$ is the transfer matrix of a two–input *MAX* gate denoted as $\mathbf{O}$. $\mathbf{T}_{\phi_2}$ is the transfer matrix representing a fanout point denoted as $\mathbf{FO}$. Stage ϕ_3 is composed of a J_1 literal selection gate in parallel with a single conducting path, thus $\mathbf{T}_{\phi_3} = \mathbf{J1} \otimes \mathbf{I}$.

The partition transfer matrices $\mathbf{T}_{\phi_i}$ as the outer product of the parallel elements within each partition and the overall monolithic transfer matrix $\mathbf{T}$ may then be computed as the direct product of the partition transfer matrices. The partition transfer matrices are

$$
\mathbf{T}_{\phi_1} = \begin{bmatrix} 1 & 0 & 0 \\ 0 & 1 & 0 \\ 0 & 0 & 1 \\ 0 & 1 & 0 \\ 0 & 1 & 0 \\ 0 & 0 & 1 \\ 0 & 0 & 1 \\ 0 & 0 & 1 \\ 0 & 0 & 1 \end{bmatrix}, \quad
\mathbf{T}_{\phi_2} = \begin{bmatrix} 1 & 0 & 0 & 0 & 0 & 0 & 0 & 0 & 0 \\ 0 & 0 & 0 & 0 & 1 & 0 & 0 & 0 & 0 \\ 0 & 0 & 0 & 0 & 0 & 0 & 0 & 0 & 1 \end{bmatrix}
$$

$$
\mathbf{T}_{\phi_3} = \begin{bmatrix} 1 & 0 & 0 \\ 0 & 0 & 1 \\ 1 & 0 & 0 \end{bmatrix} \otimes \begin{bmatrix} 1 & 0 & 0 \\ 0 & 1 & 0 \\ 0 & 0 & 1 \end{bmatrix} = \begin{bmatrix} 1 & 0 & 0 & 0 & 0 & 0 & 0 & 0 & 0 \\ 0 & 1 & 0 & 0 & 0 & 0 & 0 & 0 & 0 \\ 0 & 0 & 1 & 0 & 0 & 0 & 0 & 0 & 0 \\ 0 & 0 & 0 & 0 & 0 & 0 & 1 & 0 & 0 \\ 0 & 0 & 0 & 0 & 0 & 0 & 0 & 1 & 0 \\ 0 & 0 & 0 & 0 & 0 & 0 & 0 & 0 & 1 \\ 1 & 0 & 0 & 0 & 0 & 0 & 0 & 0 & 0 \\ 0 & 1 & 0 & 0 & 0 & 0 & 0 & 0 & 0 \\ 0 & 0 & 1 & 0 & 0 & 0 & 0 & 0 & 0 \end{bmatrix}
$$

Symbolically, the factored form of the network is expressed as

$$
\mathbf{T} = (\mathbf{O})\mathbf{FO}(\mathbf{J}_1 \otimes \mathbf{I})
$$

The explicit monolithic transfer matrix for the ternary MVSN is given in Equation 5.2.

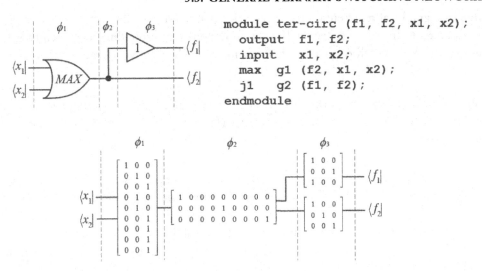

```
module ter-circ (f1, f2, x1, x2);
    output   f1, f2;
    input    x1, x2;
    max   g1 (f2, x1, x2);
    j1    g2 (f1, f2);
endmodule
```

Figure 5.6: Partitioned ternary MVSN example.

$$\mathbf{T} = \mathbf{T}_{\phi_1}\mathbf{T}_{\phi_2}\mathbf{T}_{\phi_3} = \begin{bmatrix} 1 & 0 & 0 & 0 & 0 & 0 & 0 & 0 & 0 \\ 0 & 0 & 0 & 0 & 0 & 0 & 0 & 1 & 0 \\ 0 & 0 & 1 & 0 & 0 & 0 & 0 & 0 & 0 \\ 0 & 0 & 0 & 0 & 0 & 0 & 0 & 1 & 0 \\ 0 & 0 & 0 & 0 & 0 & 0 & 0 & 1 & 0 \\ 0 & 0 & 1 & 0 & 0 & 0 & 0 & 0 & 0 \\ 0 & 0 & 1 & 0 & 0 & 0 & 0 & 0 & 0 \\ 0 & 0 & 1 & 0 & 0 & 0 & 0 & 0 & 0 \\ 0 & 0 & 1 & 0 & 0 & 0 & 0 & 0 & 0 \end{bmatrix} \tag{5.2}$$

From the properties of transfer matrices described in a previous chapter, several characteristics of the example MVSN are immediately noticeable. The rank of this 9×9 transfer matrix is 3 and the matrix contains six null column vectors $|\varnothing\varnothing)$. This observation indicates that the MVSN is only capable of producing three different distinct outputs and those output values are equivalent to the indices of the non-null column vectors as $\langle 00|$, $\langle 02|$, and $\langle 21|$.

Output Response of Ternary MVSN
As an example of how the linear algebraic framework is applied to general ternary-valued switching networks, we examine the use of the approach when used for ternary logic simulation. We utilize a netlist as input where the term "netlist" is used in the commonly accepted definition of representing a structural interconnection of logic gates. Modern EDA tools parse netlists into in-

termediate graphical representations representing a structural logic circuit and the term "netlist" does not refer to any specific format. The output response corresponding to a specified input stimulus may be computed through a direct vector-matrix multiplicative operation. This can be shown to be the case as a result of Theorem 5.3.

Lemma 5.1 Ternary Input Stimulus Linear Independence *Consider a ternary logic network with n inputs. Two distinct network input assignments $(\langle x_i|, \langle x_j|) \in \mathbb{H}^n$ are linearly independent vectors when $i \neq j$.*

Proof. $\langle x_i|$ and $\langle x_j|$ are represented as $\langle b_n b_{n-1} \dots b_0|$, where each $b_i \in \mathbb{B}$ represents a particular network input logic value. Using the fact that the input vectors can be expanded as $\langle b_n| \otimes \langle b_{n-1}| \otimes \dots \otimes \langle b_0|$. Expressed as a row vector, $\langle x| = \begin{bmatrix} 0 & 0 & \dots & 1 & \dots & 0 \end{bmatrix}$ where the single unity-valued vector component exists in a different location for $\langle x_i|$ and $\langle x_j|$ since $i \neq j$. Therefore $\langle x_i|x_j \rangle = 0$ and the norm of x_i and x_j is $L_2(x_i) = L_2(x_j) = 1$, satisfying the definition of linear independence. When $i = j$, $\langle x_i|x_j \rangle = 1$, and this only occurs when $\langle x_i| = \langle x_j|$. Expressed mathematically,

$$\mathbf{x}_i \cdot \mathbf{x}_j = \langle x_i|x_j \rangle = \begin{cases} 0, & i \neq j \\ 1, & i = j \end{cases} \tag{5.3}$$

□

Lemma 5.2 Single Ternary MVSN Output Response *The output response of a ternary MVSN due to a particular input assignment $\langle x_i|$ is represented by $\langle f_i|$. The output response vector is obtained as the product of $\langle x_i|$ with $|x_i\rangle\langle f_i|$.*

Proof. The theorem statement is expressed as

$$(\langle x_i|)(|x_i\rangle\langle f_i|) = \langle x_i|x_i\rangle\langle f_i|.$$

From Lemma 5.1, $\langle x_i|x_i\rangle = 1$, thus

$$(\langle x_i|)(|x_i\rangle\langle f_i|) = (1)\langle f_i| = \langle f_i|.$$

□

Theorem 5.3 Ternary MVSN Output Response *The transfer function representing the input-output relationship of a logic network f is of the form of a matrix $\mathbf{T}$ and is given by Equation 5.1. The output response of a ternary MVSN is obtained from the direct vector-matrix product $\langle x_i|\mathbf{T}$.*

Proof. Using Equation 5.1, the direct vector-matrix product with $\langle x_i|$ becomes

$$\langle x_i|\mathbf{T} = \langle x_i| \sum_{j=0}^{3^n-1} |x_j\rangle\langle f_j| = \sum_{j=0}^{3^n-1} \langle x_i|x_j\rangle\langle f_j|$$

From Lemma 5.1, $\langle x_i|x_j\rangle = 1$ only when $i = j$ and is otherwise zero-valued. Hence, $\langle x_i|\mathbf{T} = \langle f_i|$.

$\square$

The result of Theorem 5.3 allows for a basis of a ternary MVSN simulation technique using the vector space information model. Such a simulation can be carried output through extraction of the monolithic transfer matrix $\mathbf{T}$ followed by a direct multiplication with a specified input stimulus vector $\langle x_i|$, or through the use of a discrete event simulation approach where specified input stimuli are propagated through a graphical distributed matrix form of a netlist as depicted in the lower portion of Figure 5.3. Each approach has advantages.

The use of the monolithic transfer matrix requires that an efficient structure be used to represent the overall matrix such as a decision diagram. The multiplicative operation is then performed with a single vector representing an input stimulus in $\mathbb{H}^n$. The distributed matrix approach allows input stimuli to be specified as a collection of individual primary input assignments represented as vector in $\mathbb{H}$. The primary input values are then propagated toward the primary outputs and when network element transfer matrices are encountered, the small network element transfer matrix is multiplied with a single vector formed from the responsible internal net values. The distributed matrix approach has the advantage that only those portions of the network that contain changes in net values (or events) need be recomputed upon successive simulations. This approach is highly analogous to modern discrete event simulation methods employed in modern EDA tools that are based upon the conventional switching algebra model.

A hybrid approach can also be used for implementation of a simulation method. When the monolithic transfer matrix cannot be compactly represented, an array of transfer matrices representing each partition of a given MVSN can be formulated. This allows the transfer matrix to be represented as a factored set of serial transfer matrices. As an example of the hybrid approach, consider the example ternary MVSN depicted in Figure 5.3. Symbolically, the factored form of the transfer matrix is $\mathbf{T} = (\mathbf{O})\mathbf{F_O}(\mathbf{J}_1 \otimes \mathbf{I})$. However, this transfer matrix can also be represented symbolically in terms of the intermediate partition transfer matrices as $\mathbf{T} = \mathbf{T}_{\phi_1}\mathbf{T}_{\phi_2}\mathbf{T}_{\phi_3}$ where in this simple example, $\mathbf{T}_{\phi_1} = \mathbf{O}$, $\mathbf{T}_{\phi_2} = \mathbf{F_O}$, and $\mathbf{T}_{\phi_3} = \mathbf{J}_1 \otimes \mathbf{I}$. The simulation is then performed by computing k direct vector-matrix multiplications where k is the number of partitions or cuts since $\langle x_i|\mathbf{T} = \langle x_i|(\mathbf{T}_{\phi_1}\mathbf{T}_{\phi_2}\mathbf{T}_{\phi_3})$. The simulation can be carried out as a sequence of vector-matrix multiplications since $\langle x_i|\mathbf{T} = ((((\langle x_i|(\mathbf{T}_{\phi_1})\mathbf{T}_{\phi_2})\mathbf{T}_{\phi_3})$ thus avoiding explicit formation and storage of the monolithic transfer matrix representing the MVSN.

As r increases, the number of partially covering values increases dramatically as can be observed by the comparison of the Hasse diagrams in Figure 5.1 for $r = 2$ and $r = 3$. From observation of the form of the monolithic transfer matrix of the example ternary MVSN given in Equation 5.2, it is apparent that two possible output responses are $\langle 00|$ and $\langle 02|$. Both of these output responses can be obtained with a single simulation by specifying an input stimulus vector in the form of $\langle 0t|$ or $\langle 0t_{02}|$. To illustrate this technique, the expect calculation is carried as

$$\langle 0t|\mathbf{T} = (\langle 00| + \langle 01| + \langle 02|)\mathbf{T}$$

$$= \begin{bmatrix} 1 & 1 & 1 & 0 & 0 & 0 & 0 & 0 & 0 \end{bmatrix} \begin{bmatrix} 1 & 0 & 0 & 0 & 0 & 0 & 0 & 0 & 0 \\ 0 & 0 & 0 & 0 & 0 & 0 & 0 & 1 & 0 \\ 0 & 0 & 1 & 0 & 0 & 0 & 0 & 0 & 0 \\ 0 & 0 & 0 & 0 & 0 & 0 & 0 & 1 & 0 \\ 0 & 0 & 0 & 0 & 0 & 0 & 0 & 1 & 0 \\ 0 & 0 & 1 & 0 & 0 & 0 & 0 & 0 & 0 \\ 0 & 0 & 1 & 0 & 0 & 0 & 0 & 0 & 0 \\ 0 & 0 & 1 & 0 & 0 & 0 & 0 & 0 & 0 \\ 0 & 0 & 1 & 0 & 0 & 0 & 0 & 0 & 0 \end{bmatrix}$$

$$= \begin{bmatrix} 1 & 0 & 1 & 0 & 0 & 0 & 0 & 0 & 0 \end{bmatrix} = \langle 0t_{02}| = \langle 00| + \langle 02|.$$

Likewise, if a partial covering value such as $\langle t_{01}t_{12}|$ is specified as an input stimulus, the simulation will produce an output vector that covers two distinct output values each with a multiplicity of two as can be seen by

$$\langle t_{01}t_{12}|\mathbf{T} = (\langle 00| + \langle 01| + \langle 02|)\mathbf{T}$$

$$= \begin{bmatrix} 0 & 1 & 1 & 0 & 1 & 1 & 0 & 0 & 0 \end{bmatrix} \begin{bmatrix} 1 & 0 & 0 & 0 & 0 & 0 & 0 & 0 & 0 \\ 0 & 0 & 0 & 0 & 0 & 0 & 0 & 1 & 0 \\ 0 & 0 & 1 & 0 & 0 & 0 & 0 & 0 & 0 \\ 0 & 0 & 0 & 0 & 0 & 0 & 0 & 1 & 0 \\ 0 & 0 & 0 & 0 & 0 & 0 & 0 & 1 & 0 \\ 0 & 0 & 1 & 0 & 0 & 0 & 0 & 0 & 0 \\ 0 & 0 & 1 & 0 & 0 & 0 & 0 & 0 & 0 \\ 0 & 0 & 1 & 0 & 0 & 0 & 0 & 0 & 0 \\ 0 & 0 & 1 & 0 & 0 & 0 & 0 & 0 & 0 \end{bmatrix}$$

$$= \begin{bmatrix} 0 & 0 & 2 & 0 & 0 & 0 & 2 & 0 \end{bmatrix} = 2\begin{bmatrix} 0 & 0 & 1 & 0 & 0 & 0 & 0 & 1 & 0 \end{bmatrix}$$
$$= 2(\langle 02| + \langle 21|).$$

5.4 MVSN JUSTIFICATION

The concept of a justification matrix as previously derived for the binary radix $r = 2$ is also easily extended to the MVSN case. The multiplication is performed as $\langle f_i|\mathbf{T}^J$ where $\mathbf{T}^J$ is the "justification matrix" and $\langle f_i |$ is the output response that it is desired to justify. Furthermore, as was previously shown for the binary case $r = 2$, the justification matrix is simply the transpose of the transfer matrix, thus the complexity of performing justification is equivalent to that of the various simulation methods previously described. The pseudo-inverse of a transfer matrix for ternary MVSN can be calculated using Equation 4.2. Since the components of the transfer matrix are $t_{ij} = \mathbb{B}$, Equation 4.2 reduces to Equation refps-inverse-lab2 since $\mathbf{T}^* = \mathbf{T}^T$.

$$\mathbf{T}^+ = \begin{cases} (\mathbf{T}^T\mathbf{T})^{-1}\mathbf{T}^T, & N > M, \text{ overspecified} \\ \mathbf{T}^T(\mathbf{T}\mathbf{T}^T)^{-1}, & N < M, \text{ underspecified} \end{cases} \tag{5.4}$$

Observations from the Pseudo-inverse of T

The pseudo-inverse of a transfer function has a form that provides information regarding the characteristics of the particular MVSN it represents. The term $\mathbf{T}^T\mathbf{T}$ is the square Gram matrix, or Gramian of $\mathbf{T}$. The Gramian is useful for many other applications in linear algebra with one of those being the computation of the singular values of $\mathbf{T}$. Because transfer matrices characterizing MVSNs are in general irreversible, eigenvalues do not exist since the irreversibility occurs from the transfer matrix being non-square or not of full rank. Only for the subset of MVSNs that are reversible do eigenvalues exist for $\mathbf{T}$. In the case of irreversible transfer functions, the positive square root of the Gramian eigenvalues serve analogously to the eigenvalues of a full rank matrix.

The *Singular Value Decomposition* of a matrix allows it to be expressed as a direct product of three corresponding matrix factors, $\mathbf{U}$, $\mathbf{S}$, and $\mathbf{V}^T$. The mathematical form of the singular value decomposition of a transfer matrix $\mathbf{T}$ is then

$$\mathbf{T} = \mathbf{U}\mathbf{S}\mathbf{V}^T$$

where $\mathbf{U}$ and $\mathbf{V}^T$ are unitary matrices and $\mathbf{S}$ is the *singular value matrix*. In general, $\mathbf{T} = [t_{ij}]_{(N \times M)}$ with $N \neq N$. The dimensions of the transfer matrix are related to the number of MVSN primary inputs and outputs logarithmically since the number of primary inputs is $n = \log_r(N)$ and $m = \log_r(M)$. Since $\mathbf{U}$ and $\mathbf{V}^T$ are unitary, they are, by definition, square. The column vectors of $\mathbf{U}$ are the eigenvectors of $\mathbf{T}\mathbf{T}^T$ and the column vectors of $\mathbf{V}$ are the eigenvectors of $\mathbf{T}^T\mathbf{T}$. For the general case where $\mathbf{T}$ is non square, $\mathbf{U} = [u_{ij}]_{(N \times N)}$ and $\mathbf{V} = [v_{ij}]_{(M \times M)}$. The singular value matrix $\mathbf{S} = [s_{ij}]_{(N \times M)}$ is comprised of zero-valued components $s_{ij} = 0$ when $i \neq j$ and values $s_{ii} = \sqrt{N_i}$ along the first i diagonal components. The non-zero components of $\mathbf{S}$ are the *singular values* of $\mathbf{T}$.

The singular value decomposition is useful in relating an irreversible switching network to its reversible components since the unitary matrices $\mathbf{U}$ and $\mathbf{V}^T$ represent those portions of the overall network that are reversible. Furthermore, the Gramian has a special property for switching networks as proven for the case of a ternary MVSN in Lemma 5.4.

Theorem 5.4   Singular Values are Related to the Number of Distinct Output Responses
The singular value matrix $\mathbf{S}$ *contains diagonal singular value components that are equivalent to the number of unique output responses of the switching network characterized by* $\mathbf{T}$. *In the case of a ternary MVSN,*

$$\mathbf{S} = \begin{bmatrix} N_0 & 0 & 0 \\ 0 & N_1 & 0 \\ 0 & 0 & N_2 \end{bmatrix},$$

where N_i *is the number of or multiplicity of MVSN output responses equivalent to* $\langle (i)_{10}|$.

Proof. Any two column vectors $|t_i\rangle$ and $|t_j\rangle$, comprising the transfer matrix $\mathbf{T}$, result in $\langle t_i | t_j \rangle = 0$ when $i \neq j$ or, when $i = j$, $\langle t_i | t_i \rangle = N_i$ due to the modeling principle of using canonical basis vectors to represent each distinct switching network state of distinct output. The Gramian of $\mathbf{T} = [t_{ij}]_{(N \times M)}$ is a square $M \times M$ matrix whose column vectors are of the form $\langle t_i | t_i \rangle |i\rangle$. Thus, the Gramian of $\mathbf{T}$ is always an $M \times M$ diagonal matrix with the ith diagonal value equivalent to $\langle t_i | t_i \rangle = N_i$. The eigenvalues of a diagonal matrix are equivalent to the matrix diagonal values. By definition, the diagonal values of the singular value matrix $\mathbf{S}$ is the positive square of the eigenvalues of the Gramian. □

Corollary 5.5   Singular Values are Related to $\langle f_i | f_i \rangle$
The singular values of a transfer matrix $\mathbf{T}$ *are equivalent to* $\sqrt{\langle f_i | f_i \rangle}$ *where* $\langle f_i |$ *is the output response vector corresponding to input stimulus* $\langle i |$.

Proof. From truth table isomorphism, each row vector comprising the transfer matrix $\mathbf{T}$ is in the form of the output response arising from an input stimulus of $\langle i |$ and is thus $\langle f_i |$. Because the Gramian of $\mathbf{T}$ is of the form $\mathbf{T}^T \mathbf{T}$, each diagonal component of the Gramian is equivalent to $\langle f_i | f_i \rangle$. From the theorem, it is observed that the Gramian of $\mathbf{T}$ is always diagonal and, by definition, the singular values of $\mathbf{T}$ are the positive square roots of the eigenvalues of the Gramian $\mathbf{T}^T \mathbf{T}$. Thus, the singular values s_i, obey the relationship $s_i^2 = \langle f_i | f_i \rangle$. □

From the preceding theorem and lemmas, the pseudo-inverse of a transfer matrix $\mathbf{T}$ can be rewritten in terms of its singular value decomposition matrix $\mathbf{S}$ as

$$\mathbf{T}^+ = \begin{cases} (\mathbf{S}^2)^{-1}\mathbf{T}^T, & N > M, \text{ overspecified} \\ \mathbf{T}^T(\mathbf{S}^2)^{-1}, & N < M, \text{ underspecified} \end{cases}$$

Since S^2 is diagonal, $(S^2)^{-1}$ is easily obtained as another diagonal matrix whose diagonal components are the multiplicative inverses of the corresponding diagonal components of S. Because the singular values or diagonal components of $S = \sqrt{N_i}$, the form of S^{-1} can be expressed as given in Equation 5.5.

$$S^{-1} = \begin{bmatrix} \frac{1}{\sqrt{N_0}} & 0 & 0 & 0 & 0 & \cdots & 0 & 0 \\ 0 & \frac{1}{\sqrt{N_1}} & 0 & 0 & 0 & \cdots & 0 & 0 \\ 0 & 0 & \frac{1}{\sqrt{N_2}} & 0 & 0 & \cdots & 0 & 0 \\ 0 & 0 & 0 & \frac{1}{\sqrt{N_3}} & 0 & \cdots & 0 & 0 \\ 0 & 0 & 0 & 0 & \ddots & 0 & 0 & \vdots \\ 0 & 0 & 0 & \cdots & 0 & \frac{1}{\sqrt{N_i}} & 0 & 0 \\ 0 & 0 & 0 & \vdots & 0 & 0 & \ddots & \vdots \\ 0 & 0 & 0 & \cdots & 0 & 0 & 0 & \frac{1}{\sqrt{N_{r^{n-1}}}} \end{bmatrix} \quad (5.5)$$

In performing justification, the use of the pseudo-inverse results in weighting factors of the form of $\frac{1}{N_i}$ to be included in the resulting input responses. While the weighting information can be useful in some instances since it provides information concerning the multiplicity of common outputs for an input stimulus, it is often not needed and therefore a simpler justification matrix can be used and the inverse of the $(S^2)^{-1}$ term can be omitted from the pseudo-inverse calculation in Equation 5.4. Omission of $(S^2)^{-1}$ from the pseudo-inverse leads to the definition of the justification matrix.

Definition 5.6 *MVSN Justification Matrix*
The justification matrix T^J characterizing an MVSN is defined to be the transpose of the transfer matrix, $T^J = T^T$.

The justification matrix then provides an input stimulus that corresponds to a specified output response, but the corresponding $\frac{1}{N_i}$ weighting values are not present in the justification matrix. Mathematically, the weighting values usually do not offer any value in most EDA tasks since they normalize the output response multiplicity values so that their arithmetic sum is unity. For this reason, the use of the justification matrix in place of the pseudo-inverse is preferable not only from a computational point of view, but also due to the fact that the fractional weights in T^J are generally not useful.

$$
\begin{array}{cc|c}
a & b & MIN(a,b) \\
\hline
0 & 0 & 0 \\
0 & 1 & 0 \\
0 & 2 & 0 \\
1 & 0 & 0 \\
1 & 1 & 1 \\
1 & 2 & 1 \\
2 & 0 & 0 \\
2 & 1 & 1 \\
2 & 2 & 2 \\
\end{array}
\qquad
\mathbf{A} =
\begin{bmatrix}
1 & 0 & 0 \\
1 & 0 & 0 \\
1 & 0 & 0 \\
1 & 0 & 0 \\
0 & 1 & 0 \\
0 & 1 & 0 \\
1 & 0 & 0 \\
0 & 1 & 0 \\
0 & 0 & 1 \\
\end{bmatrix}
$$

Figure 5.7: *MIN* Gate truth table and transfer matrix.

Single Gate MVSN Justification

To illustrate justification, we consider the case of a two-input MIN gate and use both the Moore-Penrose pseudo-inverse $\mathbf{T}^+$ and the justification matrix which is simply the transpose of the transfer matrix $\mathbf{T}^J$.

Example 5.7 *Justification of Ternary MIN Gate using the Pseudo-inverse* A two-input *MIN* gate can be characterized by a truth table and corresponding isomorphic transfer matrix $\mathbf{A}$ as shown in Figure 5.7.

Using Equation 4.2, the pseudo-inverse $\mathbf{T}^+$ becomes:

$$
\mathbf{A}^+ =
\begin{bmatrix}
\frac{1}{5} & \frac{1}{5} & \frac{1}{5} & \frac{1}{5} & 0 & 0 & \frac{1}{5} & 0 & 0 \\
0 & 0 & 0 & 0 & \frac{1}{3} & \frac{1}{3} & 0 & \frac{1}{3} & 0 \\
0 & 0 & 0 & 0 & 0 & 0 & 0 & 0 & 1 \\
\end{bmatrix}
$$

Given that an output response of the *MIN*-gate is $\langle 1| = [0, 1, 0]$, the corresponding input stimulus is computed as $\langle x| = \langle f|\mathbf{A}^+$:

$$
\langle x| =
\begin{bmatrix} 0 & 1 & 0 \end{bmatrix}
\begin{bmatrix}
\frac{1}{5} & \frac{1}{5} & \frac{1}{5} & \frac{1}{5} & 0 & 0 & \frac{1}{5} & 0 & 0 \\
0 & 0 & 0 & 0 & \frac{1}{3} & \frac{1}{3} & 0 & \frac{1}{3} & 0 \\
0 & 0 & 0 & 0 & 0 & 0 & 0 & 0 & 1 \\
\end{bmatrix}
$$

$$
= \frac{1}{3} \begin{bmatrix} 0 & 0 & 0 & 0 & 1 & 1 & 0 & 1 & 0 \end{bmatrix} = \frac{1}{3}\langle 11| + \frac{1}{3}\langle 12| + \frac{1}{3}\langle 21|.
$$

Thus, it is observed that three different input stimuli can result in $\langle f| = \langle 1|$ and that those input stimuli are $\langle 11|$, $\langle 12|$, $\langle 21|$.   □

Example 5.8   *Justification of Ternary MIN Gate using* $\mathbf{A}^J$ From the principle of truth table isomorphism, the justification matrix can be obtained as the transpose of the transfer matrix shown in Figure 5.7 as

$$\mathbf{A}^J = \begin{bmatrix} 1 & 1 & 1 & 1 & 0 & 0 & 1 & 0 & 0 \\ 0 & 0 & 0 & 0 & 1 & 1 & 0 & 1 & 0 \\ 0 & 0 & 0 & 0 & 0 & 0 & 0 & 0 & 1 \end{bmatrix}$$

Given that an output response of the *MIN*-gate is $\langle 1| = [0, 1, 0]$, the corresponding input stimulus is computed as $\langle x| = \langle f|\mathbf{A}^J$:

$$\langle x| = \begin{bmatrix} 0 & 1 & 0 \end{bmatrix} \begin{bmatrix} 1 & 1 & 1 & 1 & 0 & 0 & 1 & 0 & 0 \\ 0 & 0 & 0 & 0 & 1 & 1 & 0 & 1 & 0 \\ 0 & 0 & 0 & 0 & 0 & 0 & 0 & 0 & 1 \end{bmatrix}$$

$$= \begin{bmatrix} 0 & 0 & 0 & 0 & 1 & 1 & 0 & 1 & 0 \end{bmatrix}$$
$$= \langle 11| + \langle 12| + \langle 21|$$

Thus, it is observed that three different input stimuli can result in $\langle f| = \langle 1|$ and that those input stimuli are $\langle 11|$, $\langle 12|$, and $\langle 21|$.   □

The result of the computation in Example 5.8 is an output response vector composed of nine elements. To determine the individual input stimuli, the output vector must be factored. Factoring of outer products can, in general, be complex; however due to the fact that input vector components are restricted to $\{0, 1\}$, the factors can be easily obtained by observing the indices of the non-zero components and expressing the indices in radix-3. The non-zero output response elements occur at vector indices $(4)_{10} = (11)_3$, $(5)_{10} = (12)_3$, and $(7)_{10} = (21)_3$.

To illustrate the use of other values within the Hasse diagram in Figure 5.1, consider Example 5.9.

Example 5.9   *Multiple Output Justification of Ternary MIN Gate*
Given that an output response of the *MIN*-gate is either at logic-0 or at logic-1 ($\langle f| = \langle t_{01}| = \begin{bmatrix} 1 & 1 & 0 \end{bmatrix}$), the corresponding input stimulus is computed as $\langle x| = \langle f|\mathbf{A}^+$:

$$\langle x| = \begin{bmatrix} 1 & 1 & 0 \end{bmatrix} \begin{bmatrix} \frac{1}{5} & \frac{1}{5} & \frac{1}{5} & \frac{1}{5} & 0 & 0 & \frac{1}{5} & 0 & 0 \\ 0 & 0 & 0 & 0 & \frac{1}{3} & \frac{1}{3} & 0 & \frac{1}{3} & 0 \\ 0 & 0 & 0 & 0 & 0 & 0 & 0 & 0 & 1 \end{bmatrix}$$

$$= \begin{bmatrix} \frac{1}{5} & \frac{1}{5} & \frac{1}{5} & \frac{1}{5} & \frac{1}{3} & \frac{1}{3} & \frac{1}{5} & \frac{1}{3} & 0 \end{bmatrix}$$

$$= \frac{1}{5} \begin{bmatrix} 1 & 1 & 1 & 1 & 0 & 0 & 1 & 0 & 0 \end{bmatrix} + \frac{1}{3} \begin{bmatrix} 0 & 0 & 0 & 0 & 1 & 1 & 0 & 1 & 0 \end{bmatrix}$$

$$= \frac{1}{5} (\langle 0| + \langle 1| + \langle 2| + \langle 3| + \langle 6|) + \frac{1}{3} (\langle 4| + \langle 5| + \langle 7|)$$

$$= \frac{1}{5} (\langle (00)_3| + \langle (01)_3| + \langle (02)_3| + \langle (10)_3| + \langle (20)_3|) + \frac{1}{3} (\langle (11)_3| + \langle (12)_3| + \langle (21)_3|)$$

Carrying out the justification calculation using the justification matrix $\mathbf{T}^J$ instead of the pseudo-inverse $\mathbf{T}^+$,

$$\langle x| = \begin{bmatrix} 1 & 1 & 0 \end{bmatrix} \begin{bmatrix} 1 & 1 & 1 & 1 & 0 & 0 & 1 & 0 & 0 \\ 0 & 0 & 0 & 0 & 1 & 1 & 0 & 1 & 0 \\ 0 & 0 & 0 & 0 & 0 & 0 & 0 & 0 & 1 \end{bmatrix}$$

$$= \begin{bmatrix} 1 & 1 & 1 & 1 & 1 & 1 & 1 & 1 & 0 \end{bmatrix}$$

$$= \langle 0| + \langle 1| + \langle 2| + \langle 3| + \langle 4| + \langle 5| + \langle 6| + \langle 7|$$

$$= \langle (00)_3| + \langle (01)_3| + \langle (02)_3| + \langle (10)_3| + \langle (11)_3| + \langle (12)_3| + \langle (20)_3| + \langle (21)_3|.$$

$\square$

Switching Network Justification

The justification of a switching network is an important task in many EDA applications such as various formal verification or ATPG. The use of the justification matrix allows for formulation of techniques that are identical to the simulation techniques previously described since the governing equation is of the same form. Simulation involves the solution of $\langle f_i| = \langle x_i|\mathbf{T}$ when $\langle x_i|$ and $\mathbf{T}$ are known and justification likewise requires the solution of $\langle x_i| = \langle f_i|\mathbf{T}^J$ when $\langle f_i|$ and $\mathbf{T}^J$ are known. Since $\mathbf{T}^J$ is simply $\mathbf{T}^T$, no additional netlist preprocessing is required to perform justification.

As an example of MVSN justification, we use the MVSN depicted in Figure 5.3. The justification matrix for the network is the transpose of the monolithic transfer matrix that is given in explicit matrix form in Equation 5.2.

$$\mathbf{T}^J = \mathbf{T}^T = (\mathbf{T}_{\phi_1}\mathbf{T}_{\phi_2}\mathbf{T}_{\phi_3})^T = \mathbf{T}_{\phi_3}\mathbf{T}_{\phi_2}\mathbf{T}_{\phi_1} = \begin{bmatrix} 1 & 0 & 0 & 0 & 0 & 0 & 0 & 0 & 0 \\ 0 & 0 & 0 & 0 & 0 & 0 & 0 & 0 & 0 \\ 0 & 0 & 1 & 0 & 0 & 1 & 1 & 1 & 1 \\ 0 & 0 & 0 & 0 & 0 & 0 & 0 & 0 & 0 \\ 0 & 0 & 0 & 0 & 0 & 0 & 0 & 0 & 0 \\ 0 & 0 & 0 & 0 & 0 & 0 & 0 & 0 & 0 \\ 0 & 0 & 0 & 0 & 0 & 0 & 0 & 0 & 0 \\ 0 & 1 & 0 & 1 & 1 & 0 & 0 & 0 & 0 \\ 0 & 0 & 0 & 0 & 0 & 0 & 0 & 0 & 0 \end{bmatrix} \tag{5.6}$$

According to the properties of transfer matrices as described previously, similar observations can be made with respect to the structure of the justification matrix. The justification matrix in Equation 5.6 is comprised of null row vectors at indices $\{1, 3, 4, 5, 6, 8\}$ indicating that the switching network is not capable of producing output response values of $\langle 01|$, $\langle 10|$, $\langle 11|$, $\langle 12|$, $\langle 20|$, and $\langle 22|$ where these distinct output values are simply the radix-3 form of the row indices. Furthermore, the norm of the non-null row vectors indicates the multiplicity of the number of inputs that result in the corresponding output response. The non-null row vectors at indices $\{0, 2, 7\}$ are respectively $\{1, 5, 3\}$ indicating that the output response $\langle 0|$ can occur with only a single distinct input value, $\langle 2|$ occurs for five different distinct input stimuli, and $\langle 7|$ can occur for three distinct input values.

Each column vector in $\mathbf{T}^J$ is non-null and has a unity norm due to the single non-zero component within the column vectors. The index of the non-zero column vector component specifies the input stimulus value. Examination of the structure of $\mathbf{T}^J$ results in the observation that column vectors at indices $\{2, 5, 6, 7, 8\}$ all contain a unity-valued component at index 2, therefore the justification of specified output response $\langle (2)_{10}| = \langle (02)_3|$ will yield the set of input stimuli $\langle (2)_{10}| = \langle 023|$, $\langle (5)_{10}| = \langle 123|$, $\langle (6)_{10}| = \langle 203|$, $\langle (7)_{10}| = \langle 213|$, and $\langle (8)_{10}| = \langle 223|$. Likewise, the justification of output response $\langle 7_{10}| = \langle 213|$ yields the input stimuli $\langle (1)_{10}| = \langle 013|$, $\langle (3)_{10}| = \langle 103|$, and $\langle (4)_{10}| = \langle 113|$ while output response $\langle 0_{10}| = \langle 003|$ yields the single output response $\langle 0_{10}| = \langle 003|$.

These same observations can be made without formulating the transpose of the transfer matrix by simply interchanging the words "row" and "column" in the previous two paragraphs. In essence, formulation of the monolithic transfer matrix is the only computation required to specify all simulation and justification possibilities. As will be discussed in a later chapter, it is possible to formulate simulation and justification algorithms without computing the monolithic transfer (or alternatively, the monolithic justification) matrix. While in many cases the monolithic transfer matrices can be computed and stored efficiently, alternative simulation/justification algorithms can be formulated that avoid extraction of the monolithic transfer matrices by performing simulation and justification as a traversal of the netlist. In this section, we perform justification calculations using the monolithic matrices for the purpose of illustrating various properties and characteristics of $\mathbf{T}^J$.

Example 5.10   *Justification of Output Response* $\langle 1t|$
The determination of input stimuli resulting in out responses of $\langle 10|$, $\langle 11|$, or $\langle 12|$ can be carried out by assigning $\langle f_1| = \langle 1|$ and $\langle f_2| = \langle t|$. The combined output response vector is then $\langle 1| \otimes \langle t| = \begin{bmatrix} 0 & 0 & 0 & 1 & 1 & 1 & 0 & 0 & 0 \end{bmatrix}$. By examination of the switching network in graphical form in Figure 5.3, it is apparent that none of these output responses will occur since output f_1 is produced by a J_1 literal selection gate whose only possible outputs are $\langle 0|$ or $\langle 2|$, therefore the justification calculation should produce a null input stimulus. We can verify this by performing

the following justification calculation:

$$\langle x_1 x_2| = \langle 1t|\mathbf{T}^J = \begin{bmatrix} 0 & 0 & 0 & 1 & 1 & 1 & 0 & 0 & 0 \end{bmatrix} \begin{bmatrix} 1 & 0 & 0 & 0 & 0 & 0 & 0 & 0 & 0 \\ 0 & 0 & 0 & 0 & 0 & 0 & 0 & 0 & 0 \\ 0 & 0 & 1 & 0 & 0 & 1 & 1 & 1 & 1 \\ 0 & 0 & 0 & 0 & 0 & 0 & 0 & 0 & 0 \\ 0 & 0 & 0 & 0 & 0 & 0 & 0 & 0 & 0 \\ 0 & 0 & 0 & 0 & 0 & 0 & 0 & 0 & 0 \\ 0 & 0 & 0 & 0 & 0 & 0 & 0 & 0 & 0 \\ 0 & 1 & 0 & 1 & 1 & 0 & 0 & 0 & 0 \\ 0 & 0 & 0 & 0 & 0 & 0 & 0 & 0 & 0 \end{bmatrix}$$

$$= \begin{bmatrix} 0 & 0 & 0 & 0 & 0 & 0 & 0 & 0 & 0 \end{bmatrix} = \langle \varnothing\varnothing|$$

□

It can be the case that a justification calculation will yield a non-null result when a set of output responses are justified and as few as one of the output responses are actually feasible. This is due to the fact that infeasible responses are represented by the null vector. To illustrate this case, Example 5.11 considers the case where the output responses $\langle 01|$ and $\langle 01|$ are justified.

Example 5.11 *Justification of Output Response $\langle 1t|$*
The formulation of the output response vector is obtained adding the two distinct output responses to be justified as $\langle 01| + \langle 02| = \langle 0t_{12}|$. Carrying out the justification calculation

$$\langle 0t_{12}|\mathbf{T}^J = \begin{bmatrix} 0 & 1 & 1 & 0 & 0 & 0 & 0 & 0 & 0 \end{bmatrix} \begin{bmatrix} 1 & 0 & 0 & 0 & 0 & 0 & 0 & 0 & 0 \\ 0 & 0 & 0 & 0 & 0 & 0 & 0 & 0 & 0 \\ 0 & 0 & 1 & 0 & 0 & 1 & 1 & 1 & 1 \\ 0 & 0 & 0 & 0 & 0 & 0 & 0 & 0 & 0 \\ 0 & 0 & 0 & 0 & 0 & 0 & 0 & 0 & 0 \\ 0 & 0 & 0 & 0 & 0 & 0 & 0 & 0 & 0 \\ 0 & 0 & 0 & 0 & 0 & 0 & 0 & 0 & 0 \\ 0 & 1 & 0 & 1 & 1 & 0 & 0 & 0 & 0 \\ 0 & 0 & 0 & 0 & 0 & 0 & 0 & 0 & 0 \end{bmatrix}$$

$$= \begin{bmatrix} 0 & 0 & 1 & 0 & 0 & 1 & 1 & 1 & 1 \end{bmatrix} = \langle 02| + \langle 12| + \langle 20| + \langle 21| + \langle 22|$$
$$= \langle t_{01}2| + \langle 2t|,$$

five distinct input stimuli result when the combined output response $\langle 01| + \langle 02|$ is specified. However, the output response $\langle 01|$ can never occur since $\langle 01|\mathbf{T} = \langle \varnothing\varnothing|$. Each of the five computed input stimuli will cause the switching network to produce a $\langle f_1 f_2| = \langle 02|$ output response.
□

CHAPTER 6

Binary Switching Network Spectra

6.1 SPECTRAL METHODS WITHIN THE VECTOR SPACE MODEL

Spectral methods generally refer to a class of techniques where the discrete switching functions that model a switching network are transformed to a weighted set of alternative basis functions referred to as the "spectrum" of the network. The motivation for computing various spectra is that certain EDA tasks of interest may be more efficiently carried out in the spectral versus the switching domain. This class of methods has been investigated for the past several decades and several interesting results have been obtained. In addition to numerous papers, several books have been published that describe these results [39] [38] [37] [7] [40]. Although many interesting results have been obtained, spectral methods have not been widely adopted for use in modern EDA methodology.

One of the chief reasons for this lack of adoption is the fact that the switching functions must first be extracted from a netlist or other network description in a form suitable for transformation, the transform applied, and finally the EDA task can be performed in the spectral domain. In addition to this required preprocessing before application of the EDA methodology, after it has been accomplished, an inverse transformation must be applied to return the model back into the switching function domain. Currently, these pre- and post-processing steps have overcome advantages that may be present in performing EDA tasks within the spectral domain. For this reason, considerable effort has been expended in searching for efficient methods to extract and represent both the switching functions and the spectrum and also in carrying out the forward and inverse spectral transformations.

In terms of efficient representations, some advances were obtained through formulating spectral transformations over cube list representations of switching functions [41] [42] and over decision diagram structures [43]. However, the resulting spectra remain exponentially large even when the switching functions are efficiently represented. In some cases, the spectra can also achieve some savings in representation through the use of decision diagrams, but there always remain classes of networks for which the spectra cannot be efficiently represented even when in the form of a decision diagram.

To combat this problem there have been methods developed that compute a small subset or even a single spectral coefficient, but these have enjoyed limited success [44] [45]. The technique in [44] required a structural augmentation to the netlist before each coefficient could be computed. These structural modifications resulted in large reconvergent fanout characteristics that differed for each spectral coefficient followed by a traversal of the modified circuit and carrying out a corresponding floating-point computation. In [45], the structural netlist remained unmodified; however, the resulting computed spectral coefficients were dependent upon a specific assignment to the primary inputs. In contrast, the vector space models described in this book allow for a structural netlist to remain unmodified and the resulting spectral coefficients are obtained as the spectral response in a manner analogous to obtaining the LaPlace or Fourier response of an analog circuit. The transfer matrix characterizing a structural network is transformed yielding the spectral response matrix, thus, a subset of coefficients or the entire spectrum can be computed directly from the spectral response matrix.

Another advantage of the vector space model in terms of spectral analysis, is that the netlist itself is characterized in the spectral domain rather than the output responses. Past methods rely upon transformation of switching domain output responses for spectral analysis. Here, we actually obtain a characterization of the netlist itself in the spectral domain by virtue of calculating the spectral response transfer matrix. This is analogous to common techniques in the areas of linear systems theory where systems are described in the frequency domain rather than the time domain. Should a specific spectral response be desired, the spectral transfer matrix can be evaluated for the corresponding input.

Many modern EDA methods operate over a structural netlist directly thus avoiding the intermediate extraction of a set of switching functions. The vector space model described in this book allows for the spectral description of a switching network to be extracted directly from a structural representation and furthermore for a single or small subset of spectral values to be computed from the structural representation. Therefore, both the problems of efficient representation of the network models in either the switching or spectral domains, and the problem of extraction of a single or small subset of spectral coefficients, are alleviated. The fact that these pre- and post-processing steps are now practically solved through the use of the vector space model provides motivation that spectral methods may finally enjoy more widespread usage.

There are a variety of different spectral transformations that have been used in the past. Because the spectrum is a set of weighting coefficients corresponding to a set of orthogonal basis functions, in theory any collection of orthogonal basis functions that span or cover all possible switching functions may be used. Certain collections of basis functions are popular and are named. These include the Walsh or Hadamard spectrum so named because the collection of orthogonal Walsh functions form a transformation matrix in the form of a Hadamard matrix. Another popular transformation is the Reed-Muller (RM) class of transforms. The RM transforms may be defined by any of 2^n different polarities where a given polarity corresponds to a specified set of literals in either complemented or un-complemented form.

6.2   THE SPECTRAL DOMAIN OF A SWITCHING NETWORK

From the point of view of the vector space model, we represent a specific switching network using a transfer matrix that describes the behavior of the network as a weighted sum of canonical basis vectors. For an n-input, m-output switching network, input stimuli are represented as a canonical basis vector in $\mathbb{H}^n$ and the output response vectors are weighted sums of canonical basis vectors in $\mathbb{H}^m$. The transfer matrix defines the mapping of canonical basis vectors in $\mathbb{H}^n$ to a linear combination of canonical basis vectors in $\mathbb{H}^m$. Most of the commonly used spectral transforms simply represent the output responses as linear combinations of an alternative set of basis vectors. The particular set of basis vectors used defines a particular transform. For this reason, a switching network can be characterized by any of numerous choices of mapping transformation matrices based on alternative sets of basis vectors.

6.3   FOURIER BASIS FUNCTIONS

Discrete Fourier transforms utilize basis functions composed of various sums of roots of unity along the unit circle in the complex plane. Transforms of switching functions map individual switching domain primary output responses values to points along the unit circle and then transform these to weighted sums of the basis functions defining a specific transform. As an example, Figure 6.1 contains the unit circle in the complex plane with the specific roots of unity denoted for the case of binary ($r = 2$) and ternary ($r = 3$) switching functions. In general, the form of the rth root of unity is given in Equation 6.1 where $k = 0, 1, \ldots, (r - 1)$.

$$a_k = e^{\frac{i2\pi(k)}{r}} \tag{6.1}$$

It is usually the case that the extracted switching functions representing a switching network are expressed in terms of a_i rather than the set $\mathbb{B} = \{0, 1\}$ where a_i is substituted for logic value i prior to application of a particular transform. For binary systems, $r = 2$, the roots of unity become $a_0 = 1$ and $a_1 = -1$, thus switching function values are mapped according to $\{0, 1\} \rightarrow \{+1, -1\}$. For switching networks with $r > 2$, traditional scalar-valued switching functions are mapped to complex-values corresponding to equally spaced roots of unity along the unit circle.

6.4   WALSH TRANSFORM

The Walsh transform is a Fourier transform where the basis functions correspond to the discrete Walsh functions. Walsh functions are two-valued and are thus of interest since switching functions are also two-valued. The Walsh functions can be generalized as r-valued functions leading the Chrestenson transform for switching functions of $r > 2$ values. When carrying out the Walsh transform, a particular order of the Walsh functions is predetermined and the order of the functions corresponds to variants of the Walsh transform such as the Rademacher-Walsh or the so-called natural order that leads to a transformation matrix in the form of a Hadamard matrix.

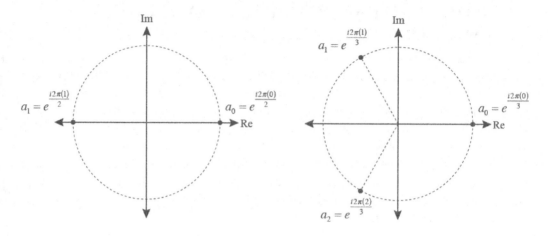

Figure 6.1: Unit circles with square and cube roots of unity.

A preliminary investigation of the Walsh transform over binary-valued switching networks when modeled in the vector space is reported in [36]. The Walsh transformation matrix when naturally ordered, becomes a Hadamard matrix. The Hadamard ordering corresponds to discretized Walsh functions where the lower-ordered functions have fewer transitions and thus allow the transformation matrices to be related for different values of n through the outer product as

$$\mathbf{H}_1 = \begin{bmatrix} 1 & 1 \\ 1 & -1 \end{bmatrix}, \quad \mathbf{H}_n = \bigotimes_{n=1}^{2^n} \mathbf{H}_1 \,.$$

6.4.1 WALSH TRANSFORM OF SCALAR-VALUED SWITCHING FUNCTIONS

Example 6.1 illustrates how the Walsh transform is computed using traditional switching functions modeled with scalars. The example uses a truth table and the transform matrix explicitly rather than the more efficient methods referenced in the previous section.

Example 6.1 *Walsh Transform Example for Scalar Switching Function*
Consider the binary-valued switching function described by the truth table. The truth table models the function values as scalar elements from $\mathbb{B}$ and also after they are mapped to roots of unity $\{a_0, a_1\}$.

x_2	x_1	x_0	f	f
0	0	0	1	-1
0	0	1	0	+1
0	1	0	1	-1
0	1	1	0	+1
1	0	0	1	-1
1	0	1	1	-1
1	1	0	0	+1
1	1	1	1	-1

To compute the Walsh spectrum of switching function f, a column vector is formed $|f\rangle$ consisting of all possible valuations followed by multiplication with the Hadamard matrix. The resulting Walsh spectrum is then a corresponding column vector $|s_f\rangle$ containing the Walsh spectral coefficients in natural (Hadamard) order. For binary-valued switching functions of n variables, the $2^n \times 2^n$ Hadamard matrix $\mathbf{H}_n$ is used to calculate the spectrum as given in Equation 6.2.

$$|s_f\rangle = \mathbf{H}_n|f\rangle \tag{6.2}$$

In this case, the switching function f depends upon three variables, thus the $\mathbf{H}_3$ Hadamard transformation is used and is calculated as:

$$\mathbf{H}_3 = \mathbf{H}_1 \otimes \mathbf{H}_1 \otimes \mathbf{H}_1 = \begin{bmatrix} 1 & 1 \\ 1 & -1 \end{bmatrix} \otimes \begin{bmatrix} 1 & 1 \\ 1 & -1 \end{bmatrix} \otimes \begin{bmatrix} 1 & 1 \\ 1 & -1 \end{bmatrix}$$

$$= \begin{bmatrix} 1 & 1 \\ 1 & -1 \end{bmatrix} \otimes \begin{bmatrix} 1 & 1 & 1 & 1 \\ 1 & -1 & 1 & -1 \\ 1 & 1 & -1 & -1 \\ 1 & -1 & -1 & 1 \end{bmatrix}$$

$$= \begin{bmatrix} 1 & 1 & 1 & 1 & 1 & 1 & 1 & 1 \\ 1 & -1 & 1 & -1 & 1 & -1 & 1 & -1 \\ 1 & 1 & -1 & -1 & 1 & 1 & -1 & -1 \\ 1 & -1 & -1 & 1 & 1 & -1 & -1 & 1 \\ 1 & 1 & 1 & 1 & -1 & -1 & -1 & -1 \\ 1 & -1 & 1 & -1 & -1 & 1 & -1 & 1 \\ 1 & 1 & -1 & -1 & -1 & -1 & 1 & 1 \\ 1 & -1 & -1 & 1 & -1 & 1 & 1 & -1 \end{bmatrix}.$$

Each component of $|s_f\rangle$ is the spectral coefficient corresponding to a particular variable assignment of f. When f represents a switching network, each spectral coefficient can be considered as a spectral output response of a network.

$$|s_f\rangle = \mathbf{H}_3|f\rangle = \begin{bmatrix} 1 & 1 & 1 & 1 & 1 & 1 & 1 & 1 \\ 1 & -1 & 1 & -1 & 1 & -1 & 1 & -1 \\ 1 & 1 & -1 & -1 & 1 & 1 & -1 & -1 \\ 1 & -1 & -1 & 1 & 1 & -1 & -1 & 1 \\ 1 & 1 & 1 & 1 & -1 & -1 & -1 & -1 \\ 1 & -1 & 1 & -1 & -1 & 1 & -1 & 1 \\ 1 & 1 & -1 & -1 & -1 & -1 & 1 & 1 \\ 1 & -1 & -1 & 1 & -1 & 1 & 1 & -1 \end{bmatrix} \begin{bmatrix} -1 \\ +1 \\ -1 \\ +1 \\ -1 \\ -1 \\ +1 \\ -1 \end{bmatrix} = \begin{bmatrix} -2 \\ -2 \\ -2 \\ -2 \\ +2 \\ -6 \\ +2 \\ +2 \end{bmatrix}$$

□

The Walsh spectrum has many properties of interest. When the matrix components are scaled by the value $\frac{1}{2^n}$, a unitary matrix results, thus the inverse transform can be computed with same matrix. Since the convention is not to scale the transform matrix with $\frac{1}{2^n}$, the inverse is computed by the inverse transform $\frac{1}{2^n}\mathbf{H}_n$.

Early attempts to optimize the computation of this class of Fourier transforms resulted in the so-called "fast" transforms. Fast transforms can be viewed as applying the transformation matrix in factored form, that is, as a set of outer product factors rather than the overall monolithic transformation matrix. This is expressed mathematically as

$$|s_f\rangle = \left(\bigotimes_{n=1}^{2^n} \mathbf{H}_1\right)|f\rangle.$$

In the literature, the fast transforms are commonly depicted in a graphical form using a signal flow graph referred to as a "butterfly" diagram [46]. The signal flow graph or butterfly diagram depicts the transform where directed edges represent multiplication with value from the factored form of the transformation matrix and the vertices represent an addition operation. The term "butterfly" arises from the shape of the signal flow graph for the $\mathbf{H}_1$ transformation matrix. Figure 6.2 depicts the signal flow graph for $\mathbf{H}_1$ and Figure 6.3 depicts the signal flow graph for $\mathbf{H}_3$ and is annotated with the values from the specific example in Example 6.1. For the case $r = 2$, all matrix coefficients are ±1, thus a shorthand notation is to represent -1 with a dashed line and +1 with a solid line in the butterfly diagram.

The development of the class of "fast" transforms was notable since it allowed for any Fourier-like transform to be implemented in a computer algorithm requiring fewer arithmetic operations than direct implementation of Equation 6.2 using the monolithic transfer matrix. The formulation of the "fast" Fourier transform is usually credited to the work in [47] [48] although there are indications that this shorthand method was used earlier by other mathematicians for various applications.

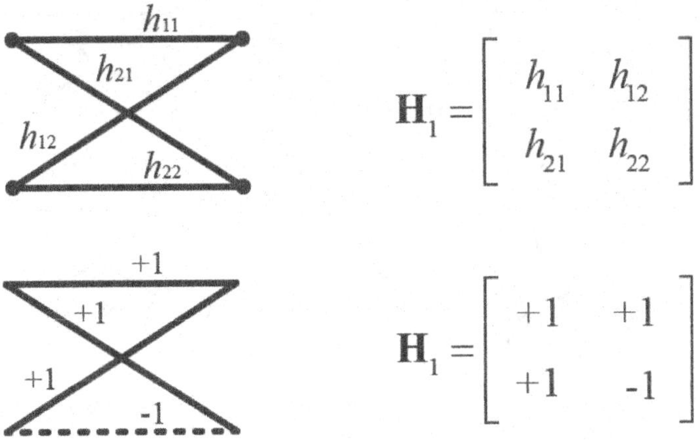

$$\mathbf{H}_1 = \begin{bmatrix} h_{11} & h_{12} \\ h_{21} & h_{22} \end{bmatrix}$$

$$\mathbf{H}_1 = \begin{bmatrix} +1 & +1 \\ +1 & -1 \end{bmatrix}$$

Figure 6.2: $\mathbf{H}_1$ and corresponding butterfly diagrams.

The concept of the fast transform can be applied to representations of the vectors $|f\rangle$ and $|s_f\rangle$ in other forms leading to increased efficiency. One example is the use of the decision diagram structure allowing for the fast transform to be implemented as an algorithm operating on directed graphs. Although variants of this approach have been formulated, the basic idea is to traverse a BDD representing $|f\rangle$ and to apply the butterfly operation to each BDD vertex. The resulting decision diagram then represents the vector $|s_f\rangle$. This type of approach is described in detail for the Walsh and other types of transformation in [43] [49] [7] [37]. While the use of decision diagrams did allow for the practical computation of spectra for underlying switching circuits than was previously possible, limitations still existed in that the initial extraction of a switching function and formulating the BDD representation was required and in some cases, the BDD size remained exponential. Furthermore, some classes of switching functions with compact BDD representations in the switching domain grew to be unacceptably large when transformed into a DD representing the spectrum.

The use of cube lists to compute spectra was also developed in an attempt to allow the computations to be performed more efficiently. However, compact cube lists generally contain many instances where two or more cubes cover the same function values. This covering set of cubes can then lead to complications in computing spectral coefficients since the joint covering phenomena must be accounted for. If the cube list is converted to a disjoint cover, the problem of the list growing in size occurs in a similar manner to the problem that some BDDs are very large.

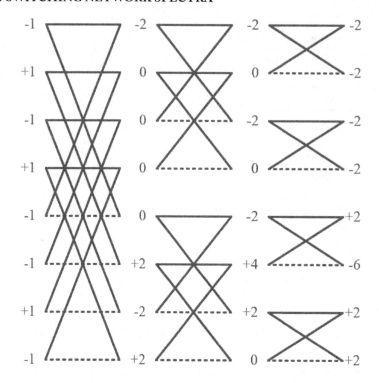

Figure 6.3: Butterfly diagram for Example 6.1 calculation.

6.4.2   WALSH TRANSFORM FOR VECTOR-VALUED SWITCHING FUNCTIONS

When a switching network is modeled in the vector space, the Walsh transform can also be applied as in the previous section. For the $r = 2$ binary case, the values $\langle 0|$ and $\langle 1|$ can be mapped to values along the unit circle in an analogous manner to that of the scalar-valued case. This mapping is obtained by applying the $\mathbf{H_1}$ transformation to each value so that it may be represented in the spectral domain.

$$\langle 0|\mathbf{H_1} = \begin{bmatrix} 1 & 0 \end{bmatrix} \begin{bmatrix} 1 & 1 \\ 1 & -1 \end{bmatrix} = \begin{bmatrix} 1 & 1 \end{bmatrix} \quad \langle 1|\mathbf{H_1} = \begin{bmatrix} 0 & 1 \end{bmatrix} \begin{bmatrix} 1 & 1 \\ 1 & -1 \end{bmatrix} = \begin{bmatrix} 1 & -1 \end{bmatrix}$$

The vector space model also allows for the values $\langle t|$ and $\langle \varnothing|$. These values in the Walsh domain are expressed as follows.

$$\langle\varnothing|\mathbf{H}_1 = \begin{bmatrix} 0 & 0 \end{bmatrix}\begin{bmatrix} 1 & 1 \\ 1 & -1 \end{bmatrix} = \begin{bmatrix} 0 & 0 \end{bmatrix} \quad \langle t|\mathbf{H}_1 = \begin{bmatrix} 1 & 1 \end{bmatrix}\begin{bmatrix} 1 & 1 \\ 1 & -1 \end{bmatrix} = \begin{bmatrix} 2 & 0 \end{bmatrix}$$

In this manner, the Walsh transform matrix $\mathbf{H}_1$ can be viewed as the collection of row vectors representing the spectral values corresponding to $\langle 0|$ and $\langle 1|$. The Hasse diagram of the vector-valued constants in the switching domain and the Walsh spectral domain are given in Figure 6.4.

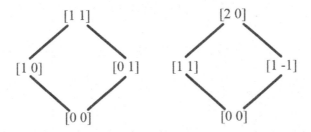

Figure 6.4: Hasse diagram of values in the switching and Walsh spectral domain for $r = 2$.

The Walsh spectrum of a switching network when modeled in the vector space can analogously be computed as in the case of the more conventional scalar model described in the previous section. A column vector of all possible output values can be formulated and multiplied with the Walsh transform matrix. However, each component of the column vector is also a row vector since a row vector corresponds to each distinct output response. Therefore, the column vector containing all network output responses is itself a matrix, and this matrix is exactly the transfer matrix $\mathbf{T}_f$ characterizing the switching network. The transfer matrix row vectors are mapped to corresponding values as given in the Hasse diagram in Figure 6.4 for consistency with the definition.

Theorem 6.2   Walsh Spectrum of Switching Network
The Walsh spectrum of switching network modeled in the vector space is equivalent to the product of the mapped transfer matrix $\mathbf{T}_s$ and the Walsh transformation matrix $\mathbf{H}_n$.

Proof. By definition, the spectrum of a switching function is the product of the Walsh transformation matrix with a column vector whose components are all possible function values. By the property of truth table isomorphism, the transfer matrix characterizing a switching network consists of row vectors that are all possible output responses of the network. By the definition of the spectrum of network, the spectrum of the network is

$$\mathbf{S}_f = \mathbf{H}_n \mathbf{T}_s.$$

□

Example 6.3   *Walsh Spectrum of Example Function*
Consider the single output switching characterized by the truth table that contains both the scalar model and vector model values.

x_2	x_1	x_0	f	$\langle f |$
0	0	0	1	$\begin{bmatrix} 0 & 1 \end{bmatrix}$
0	0	1	0	$\begin{bmatrix} 1 & 0 \end{bmatrix}$
0	1	0	1	$\begin{bmatrix} 0 & 1 \end{bmatrix}$
0	1	1	0	$\begin{bmatrix} 1 & 0 \end{bmatrix}$
1	0	0	1	$\begin{bmatrix} 0 & 1 \end{bmatrix}$
1	0	1	1	$\begin{bmatrix} 0 & 1 \end{bmatrix}$
1	1	0	0	$\begin{bmatrix} 1 & 0 \end{bmatrix}$
1	1	1	1	$\begin{bmatrix} 0 & 1 \end{bmatrix}$

From the definition of the Walsh spectrum, a column vector is formed with components equivalent to the truth table values of the function $\langle f |$. This column vector is equivalent to the transfer matrix characterizing $\langle f |$.

$$\mathbf{T}_f = \begin{bmatrix} 0 & 1 \\ 1 & 0 \\ 0 & 1 \\ 1 & 0 \\ 0 & 1 \\ 0 & 1 \\ 1 & 0 \\ 0 & 1 \end{bmatrix}$$

The mapped version of the transfer matrix is denoted as $\mathbf{T}_s$:

$$
\mathbf{T}_s = \begin{bmatrix} 1 & -1 \\ 1 & 1 \\ 1 & -1 \\ 1 & 1 \\ 1 & -1 \\ 1 & -1 \\ 1 & 1 \\ 1 & -1 \end{bmatrix}
$$

The Walsh spectrum is then computed as $\mathbf{S}_f = \mathbf{H}_f \mathbf{T}_s$:

$$
\mathbf{S}_f = \begin{bmatrix}
1 & 1 & 1 & 1 & 1 & 1 & 1 & 1 \\
1 & -1 & 1 & -1 & 1 & -1 & 1 & -1 \\
1 & 1 & -1 & -1 & 1 & 1 & -1 & -1 \\
1 & -1 & -1 & 1 & 1 & -1 & -1 & 1 \\
1 & 1 & 1 & 1 & -1 & -1 & -1 & -1 \\
1 & -1 & 1 & -1 & -1 & 1 & -1 & 1 \\
1 & 1 & -1 & -1 & -1 & -1 & 1 & 1 \\
1 & -1 & -1 & 1 & -1 & 1 & 1 & -1
\end{bmatrix}
\begin{bmatrix}
1 & -1 \\
1 & 1 \\
1 & -1 \\
1 & 1 \\
1 & -1 \\
1 & -1 \\
1 & 1 \\
1 & -1
\end{bmatrix}
=
\begin{bmatrix}
8 & -2 \\
0 & -2 \\
0 & -2 \\
0 & -2 \\
0 & +2 \\
0 & -6 \\
0 & +2 \\
0 & +2
\end{bmatrix}
$$

Theorem 6.2 leads to the following definition.

Definition 6.4   *Spectral Response Matrix*
The spectral response matrix $\mathbf{T}_s$ characterizes a switching network in the spectral domain and is equivalent to the transfer matrix $\mathbf{T}$ with each row vector mapped to their corresponding spectral representations.

Each spectral coefficient can be calculated individually through a multiplicative operation with a specific row vector. The row vector represents a switching network input stimulus expressed where each individual primary input value is expressed in the spectral domain. If the input stimulus is represented in the switching domain, the spectral response matrix yields a corresponding output response with each primary output expressed in the spectral domain. Example 6.5 illustrates these two uses of the spectral response matrix.

Example 6.5   *Spectral Response Matrix Calculations*
Consider the example single-output switching network characterized by the following truth table.

x_2	x_1	x_0	f	$\langle f \vert$	$\langle f_s \vert$
0	0	0	1	$\begin{bmatrix} 0 & 1 \end{bmatrix}$	$\begin{bmatrix} 1 & -1 \end{bmatrix}$
0	0	1	0	$\begin{bmatrix} 1 & 0 \end{bmatrix}$	$\begin{bmatrix} 1 & 1 \end{bmatrix}$
0	1	0	1	$\begin{bmatrix} 0 & 1 \end{bmatrix}$	$\begin{bmatrix} 1 & -1 \end{bmatrix}$
0	1	1	0	$\begin{bmatrix} 1 & 0 \end{bmatrix}$	$\begin{bmatrix} 1 & 1 \end{bmatrix}$
1	0	0	1	$\begin{bmatrix} 0 & 1 \end{bmatrix}$	$\begin{bmatrix} 1 & -1 \end{bmatrix}$
1	0	1	1	$\begin{bmatrix} 0 & 1 \end{bmatrix}$	$\begin{bmatrix} 1 & -1 \end{bmatrix}$
1	1	0	0	$\begin{bmatrix} 1 & 0 \end{bmatrix}$	$\begin{bmatrix} 1 & 1 \end{bmatrix}$
1	1	1	1	$\begin{bmatrix} 0 & 1 \end{bmatrix}$	$\begin{bmatrix} 1 & -1 \end{bmatrix}$

The spectral response matrix is then:

$$\mathbf{T}_s = \begin{bmatrix} 1 & -1 \\ 1 & 1 \\ 1 & -1 \\ 1 & 1 \\ 1 & -1 \\ 1 & -1 \\ 1 & 1 \\ 1 & -1 \end{bmatrix}$$

Spectral coefficients may be calculated individually by forming a row vector corresponding to each spectral coefficient and multiplying with the spectral response matrix. For instance, the spectral coefficient corresponding to input stimulus $\langle 011 \vert$ is computed by first mapping each distinct input to its spectral representation, then multiplying it with the spectral response matrix. Mapping the input stimulus vector:

$$\langle 011 \vert = \begin{bmatrix} 1 & 0 \end{bmatrix} \otimes \begin{bmatrix} 0 & 1 \end{bmatrix} \otimes \begin{bmatrix} 0 & 1 \end{bmatrix} \rightarrow \begin{bmatrix} 1 & 1 \end{bmatrix} \otimes \begin{bmatrix} 1 & -1 \end{bmatrix} \otimes \begin{bmatrix} 1 & -1 \end{bmatrix}$$
$$= \begin{bmatrix} 1 & 1 \end{bmatrix} \otimes \begin{bmatrix} 1 & -1 & 1 & -1 \end{bmatrix} = \begin{bmatrix} 1 & -1 & 1 & -1 & 1 & -1 & 1 & -1 \end{bmatrix}$$

The spectral coefficient is then computed as:

$$\langle s \vert_{011} = \begin{bmatrix} 1 & -1 & 1 & -1 & 1 & -1 & 1 & -1 \end{bmatrix} \begin{bmatrix} 1 & -1 \\ 1 & 1 \\ 1 & -1 \\ 1 & 1 \\ 1 & -1 \\ 1 & -1 \\ 1 & 1 \\ 1 & -1 \end{bmatrix} = \begin{bmatrix} 0 & -2 \end{bmatrix}$$

Alternatively, the spectral response matrix can be used to determine the output response due to input stimulus $\langle 011|$ as follows.

$$\langle 011| = \begin{bmatrix} 0 & 0 & 0 & 1 & 0 & 0 & 0 & 0 \end{bmatrix} \begin{bmatrix} 1 & -1 \\ 1 & 1 \\ 1 & -1 \\ 1 & 1 \\ 1 & -1 \\ 1 & -1 \\ 1 & 1 \\ 1 & -1 \end{bmatrix} = \begin{bmatrix} 1 & 1 \end{bmatrix}$$

The result $\begin{bmatrix} 1 & 1 \end{bmatrix}$ is the value of $\langle 0|$ in the Walsh spectrum as can be verified by taking the inverse transform.

$$\begin{bmatrix} 1 & 1 \end{bmatrix} \frac{1}{2} \begin{bmatrix} 1 & 1 \\ 1 & -1 \end{bmatrix} = \frac{1}{2} \begin{bmatrix} 2 & 0 \end{bmatrix} = \begin{bmatrix} 1 & 0 \end{bmatrix} = \langle 0|$$

□

This spectral response matrix may also be used to compute subsets or even the entire spectrum through multiplication by matrices whose row space varies.

Lemma 6.6 Partial Walsh Spectrum Using Spectral Response Matrix
The partial Walsh spectrum consisting of $k < 2^n$ coefficients for a switching network characterized by a spectral transfer matrix $\mathbf{T}_s$ is given by $\mathbf{V}\mathbf{T}_s$. Where $\mathbf{V} = [v_{ij}]_{k \times 2^n}$ and each row vector in $\mathbf{V}$ corresponds to a distinct spectral coefficient.

Proof. The matrix characterizing all possible input stimuli is $\mathbf{I}$ where each row vector of the identity matrix represents a canonical basis vector that models a specific input stimulus. By Theorem 6.2, the complete spectrum is given by $\mathbf{H}_n \mathbf{T}_s$, and the Walsh spectral matrix arises because the mapping of row vectors in $\mathbf{I}$ corresponds to $\mathbf{I}\mathbf{H}_n = \mathbf{H}_n$. Therefore, a subset of coefficients may be computed by forming the matrix $\mathbf{V}$ comprised of k of row vectors from $\mathbf{H}_n$. □

Example 6.7 *Computing a Subset of Spectral Coefficients*
Consider the example switching function characterized by the truth table:

x_2	x_1	x_0	f	$\langle f|$	$\langle f_s|$
0	0	0	1	$\begin{bmatrix} 0 & 1 \end{bmatrix}$	$\begin{bmatrix} 1 & -1 \end{bmatrix}$
0	0	1	0	$\begin{bmatrix} 1 & 0 \end{bmatrix}$	$\begin{bmatrix} 1 & 1 \end{bmatrix}$
0	1	0	1	$\begin{bmatrix} 0 & 1 \end{bmatrix}$	$\begin{bmatrix} 1 & -1 \end{bmatrix}$
0	1	1	0	$\begin{bmatrix} 1 & 0 \end{bmatrix}$	$\begin{bmatrix} 1 & 1 \end{bmatrix}$
1	0	0	1	$\begin{bmatrix} 0 & 1 \end{bmatrix}$	$\begin{bmatrix} 1 & -1 \end{bmatrix}$
1	0	1	1	$\begin{bmatrix} 0 & 1 \end{bmatrix}$	$\begin{bmatrix} 1 & -1 \end{bmatrix}$
1	1	0	0	$\begin{bmatrix} 1 & 0 \end{bmatrix}$	$\begin{bmatrix} 1 & 1 \end{bmatrix}$
1	1	1	1	$\begin{bmatrix} 0 & 1 \end{bmatrix}$	$\begin{bmatrix} 1 & -1 \end{bmatrix}$

The complete spectrum of this example function is a matrix $\mathbf{S}_f$ whose row vectors are distinct spectral coefficients. From this point of view, the spectrum may be written as:

$$
\mathbf{S}_f = \begin{bmatrix}
\langle s_{000}| \\
\langle s_{001}| \\
\langle s_{010}| \\
\langle s_{011}| \\
\langle s_{100}| \\
\langle s_{101}| \\
\langle s_{110}| \\
\langle s_{111}|
\end{bmatrix}
$$

A subset of spectral coefficients can be calculated by forming a matrix $\mathbf{V}$ consisting of a set of canonical row vectors expressed in their Walsh spectral form. Assume it is desired to calculate $\langle s_{001}|$, $\langle s_{011}|$, $\langle s_{101}|$, and $\langle s_{111}|$. The matrix $\mathbf{V}$ is:

$$
\mathbf{V} =
\begin{bmatrix}
0 & 1 & 0 & 0 & 0 & 0 & 0 & 0 \\
0 & 0 & 0 & 1 & 0 & 0 & 0 & 0 \\
0 & 0 & 0 & 0 & 0 & 1 & 0 & 0 \\
0 & 0 & 0 & 0 & 0 & 0 & 0 & 1
\end{bmatrix}
\begin{bmatrix}
1 & 1 & 1 & 1 & 1 & 1 & 1 & 1 \\
1 & -1 & 1 & -1 & 1 & -1 & 1 & -1 \\
1 & 1 & -1 & -1 & 1 & 1 & -1 & -1 \\
1 & -1 & -1 & 1 & 1 & -1 & -1 & 1 \\
1 & 1 & 1 & 1 & -1 & -1 & -1 & -1 \\
1 & -1 & 1 & -1 & -1 & 1 & -1 & 1 \\
1 & 1 & -1 & -1 & -1 & -1 & 1 & 1 \\
1 & -1 & -1 & 1 & -1 & 1 & 1 & -1
\end{bmatrix}
$$

$$
=
\begin{bmatrix}
1 & -1 & 1 & -1 & 1 & -1 & 1 & -1 \\
1 & -1 & -1 & 1 & 1 & -1 & -1 & 1 \\
1 & -1 & 1 & -1 & -1 & 1 & -1 & 1 \\
1 & -1 & -1 & 1 & -1 & 1 & 1 & -1
\end{bmatrix}
$$

Notice that $\mathbf{V}$ is simply a subset of row vectors from the spectral transformation matrix $\mathbf{H}_n$. The desired subset of spectral coefficients is computed as:

$$\mathbf{VT}_s = \begin{bmatrix} 1 & -1 & 1 & -1 & 1 & -1 & 1 & -1 \\ 1 & -1 & -1 & 1 & 1 & -1 & -1 & 1 \\ 1 & -1 & 1 & -1 & -1 & 1 & -1 & 1 \\ 1 & -1 & -1 & 1 & -1 & 1 & 1 & -1 \end{bmatrix} \begin{bmatrix} 1 & -1 \\ 1 & 1 \\ 1 & -1 \\ 1 & 1 \\ 1 & -1 \\ 1 & -1 \\ 1 & 1 \\ 1 & -1 \end{bmatrix} = \begin{bmatrix} 0 & -2 \\ 0 & -2 \\ 0 & -6 \\ 0 & 2 \end{bmatrix}$$

$\square$

It is also possible to compute a sum of two or more spectral coefficients through vector addition of the row vectors specifying the spectral coefficients. The following example illustrates this calculation.

Example 6.8   *Sum of Spectral Coefficients*
Consider the example function specified in truth table form in Example 6.7. In that previous example, the distinct spectral coefficients $\langle s_{001}|$, $\langle s_{011}|$, $\langle s_{101}|$, and $\langle s_{111}|$ were calculated. Here, we calculate the sum of these four spectral coefficients by forming a single row vector whose components are the sum of the corresponding components in the row vectors of $\mathbf{V}$ and multiply the vector with the spectral response matrix. The sum vector, $\langle v_{sum}|$ is

$$\langle v_{sum}| = \begin{bmatrix} 4 & -4 & 0 & 0 & 0 & 0 & 0 & 0 \end{bmatrix}.$$

Multiplying this vector with the spectral response matrix:

$$\begin{bmatrix} 4 & -4 & 0 & 0 & 0 & 0 & 0 & 0 \end{bmatrix} \begin{bmatrix} 1 & -1 \\ 1 & 1 \\ 1 & -1 \\ 1 & 1 \\ 1 & -1 \\ 1 & -1 \\ 1 & 1 \\ 1 & -1 \end{bmatrix} = \begin{bmatrix} 0 & -8 \end{bmatrix}$$

It can be verified that this result is indeed the sum of the spectral coefficients $\langle s_{001}|$, $\langle s_{011}|$, $\langle s_{101}|$, and $\langle s_{111}|$ by directly adding them together.

$$\begin{bmatrix} 0 & -2 \end{bmatrix} + \begin{bmatrix} 0 & -2 \end{bmatrix} + \begin{bmatrix} 0 & -6 \end{bmatrix} + \begin{bmatrix} 0 & 2 \end{bmatrix} = \begin{bmatrix} 0 & -8 \end{bmatrix}$$

□

6.4.3 WALSH SPECTRAL RESPONSE MATRICES

Multi-output switching networks are characterized with transfer matrices that yield output responses in the form of the outer product of each individual primary output response. In a similar manner, spectral responses are also in the form of outer products of their individual primary outputs. To describe this characteristic, we use the example switching network in Figure 6.5.

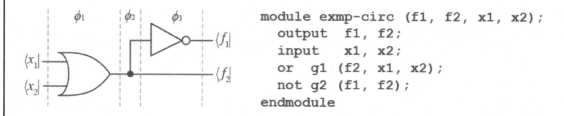

```
module exmp-circ (f1, f2, x1, x2);
    output   f1, f2;
    input    x1, x2;
    or   g1 (f2, x1, x2);
    not  g2 (f1, f2);
endmodule
```

Figure 6.5: Example logic network HDL and diagram.

This netlist has a characterizing transfer matrix $\mathbf{T}$ and the corresponding spectral response matrix $\mathbf{T}_s$ is computed as follows. First, the transfer matrix is written in a form where the row vectors are expressed as outer product factors.

$$\mathbf{T} = \begin{bmatrix} 0 & 0 & 1 & 0 \\ 0 & 1 & 0 & 0 \\ 0 & 1 & 0 & 0 \\ 0 & 1 & 0 & 0 \end{bmatrix} = \begin{bmatrix} \langle 10| \\ \langle 01| \\ \langle 01| \\ \langle 01| \end{bmatrix} = \begin{bmatrix} \begin{bmatrix} 0 & 1 \end{bmatrix} \otimes \begin{bmatrix} 1 & 0 \end{bmatrix} \\ \begin{bmatrix} 1 & 0 \end{bmatrix} \otimes \begin{bmatrix} 0 & 1 \end{bmatrix} \\ \begin{bmatrix} 1 & 0 \end{bmatrix} \otimes \begin{bmatrix} 0 & 1 \end{bmatrix} \\ \begin{bmatrix} 1 & 0 \end{bmatrix} \otimes \begin{bmatrix} 0 & 1 \end{bmatrix} \end{bmatrix}$$

Following the definition of the spectral response matrix, each instance of $\langle 0|$ and $\langle 1|$ is replaced with its Walsh spectral equivalents given by $\langle 0|\mathbf{H}_1$ and $\langle 1|\mathbf{H}_1$.

$$\mathbf{T}_s = \begin{bmatrix} \begin{bmatrix} 0 & 1 \end{bmatrix} \mathbf{H}_1 \otimes \begin{bmatrix} 1 & 0 \end{bmatrix} \mathbf{H}_1 \\ \begin{bmatrix} 1 & 0 \end{bmatrix} \mathbf{H}_1 \otimes \begin{bmatrix} 0 & 1 \end{bmatrix} \mathbf{H}_1 \\ \begin{bmatrix} 1 & 0 \end{bmatrix} \mathbf{H}_1 \otimes \begin{bmatrix} 0 & 1 \end{bmatrix} \mathbf{H}_1 \\ \begin{bmatrix} 1 & 0 \end{bmatrix} \mathbf{H}_1 \otimes \begin{bmatrix} 0 & 1 \end{bmatrix} \mathbf{H}_1 \end{bmatrix}$$

Using the outer product property $\mathbf{AC} \otimes \mathbf{BC} = (\mathbf{A} \otimes \mathbf{B})(\mathbf{C} \otimes \mathbf{C})$, the expression for the spectral transfer matrix becomes:

$$
\mathbf{T}_s = \begin{bmatrix} (\begin{bmatrix} 0 & 1 \end{bmatrix} \otimes \begin{bmatrix} 1 & 0 \end{bmatrix})\mathbf{H}_2 \\ (\begin{bmatrix} 1 & 0 \end{bmatrix} \otimes \begin{bmatrix} 0 & 1 \end{bmatrix})\mathbf{H}_2 \\ (\begin{bmatrix} 1 & 0 \end{bmatrix} \otimes \begin{bmatrix} 0 & 1 \end{bmatrix})\mathbf{H}_2 \\ (\begin{bmatrix} 1 & 0 \end{bmatrix} \otimes \begin{bmatrix} 0 & 1 \end{bmatrix})\mathbf{H}_2 \end{bmatrix} = \begin{bmatrix} \langle 10| \\ \langle 01| \\ \langle 01| \\ \langle 01| \end{bmatrix} \mathbf{H}_2 = \mathbf{T}\mathbf{H}_2.
$$

Therefore the spectral response matrix characterizing a switching network of m primary outputs can be calculated by multiplying the transfer matrix with the Walsh transformation matrix, $\mathbf{H}_m$. Carrying out this computation for the example netlist in Figure 6.5,

$$
\mathbf{T}_s = \begin{bmatrix} 0 & 0 & 1 & 0 \\ 0 & 1 & 0 & 0 \\ 0 & 1 & 0 & 0 \\ 0 & 1 & 0 & 0 \end{bmatrix} \begin{bmatrix} 1 & 1 & 1 & 1 \\ 1 & -1 & 1 & -1 \\ 1 & 1 & -1 & -1 \\ 1 & -1 & -1 & 1 \end{bmatrix} = \begin{bmatrix} 1 & 1 & -1 & -1 \\ 1 & -1 & 1 & -1 \\ 1 & -1 & 1 & -1 \\ 1 & -1 & 1 & -1 \end{bmatrix}
$$

The spectral response matrix fully characterizes the example netlist and can be used to compute a subset of or the entire Walsh spectrum, or to compute the output response in the spectral domain. For this reason, the spectral response matrix is analogous to the transfer function of a continuous linear system in the Fourier domain. The following two examples illustrate how the spectral response matrix can be used to carry out these computations.

Example 6.9   *Calculation of a Walsh Spectral Coefficient for Example Netlist*
Assume the spectral coefficient $\langle s_{10}|$ is desired. This coefficient can be computed as follows. First, the appropriate input stimulus row vector is formed by multiplying $\langle 10|\mathbf{H}_2$. Notice, this is actually the third row vector from the top of $\mathbf{H}_2$ since $\langle 10|$ is a canonical basis vector in $\mathbb{H}^2$. Next, this row vector is multiplied with the network spectral response matrix yielding the following.

$$
\begin{bmatrix} 1 & 1 & -1 & -1 \end{bmatrix} \begin{bmatrix} 1 & 1 & -1 & -1 \\ 1 & -1 & 1 & -1 \\ 1 & -1 & 1 & -1 \\ 1 & -1 & 1 & -1 \end{bmatrix} = \begin{bmatrix} 0 & 2 & -2 & 0 \end{bmatrix}
$$

The resultant spectral coefficient vector is in the form of a row vector in $\mathbb{H}^2$ since it corresponds to each of the two network primary outputs.   □

The spectral response matrix may also be used to determine the output response due to a specific input stimulus as shown in Example 6.10.

Example 6.10   *Calculation of an Output Response for Example Netlist*
If it is desired to compute the output response for $\langle x_1 x_2| = \langle 10|$ with the spectral response matrix

instead of the transfer matrix, the calculation is very similar to that of the preceding Example 6.9 with the exception that the canonical basis vector $\langle 10|$ is not specified in the Walsh spectral domain. Because the spectral response matrix is being used, the corresponding output response will be in the Walsh domain denoted as $\langle f_{s1} f_{s2}|$, but can easily be converted back to the switching domain yielding $\langle f_1 f_2|$ by multiplying with the inverse Walsh transformation. The following calculation illustrates this computation.

$$\langle f_{s1} f_{s2}| = \langle 10|\mathbf{T}_s = \begin{bmatrix} 0 & 0 & 1 & 0 \end{bmatrix} \begin{bmatrix} 1 & 1 & -1 & -1 \\ 1 & -1 & 1 & -1 \\ 1 & -1 & 1 & -1 \\ 1 & -1 & 1 & -1 \end{bmatrix} = \begin{bmatrix} 1 & -1 & 1 & -1 \end{bmatrix}$$

The resultant output response is specified in the Walsh spectral domain and can be converted back to the switching domain by computing $\langle f_{s1} f_{s2}|(\frac{1}{4}\mathbf{H}_2)$ as shown in the following calculation.

$$\langle f_1 f_2| = \langle f_{s1} f_{s2}|\mathbf{H}_2 = \begin{bmatrix} 1 & -1 & 1 & -1 \end{bmatrix} \left(\frac{1}{4} \begin{bmatrix} 1 & 1 & 1 & 1 \\ 1 & -1 & 1 & -1 \\ 1 & 1 & -1 & -1 \\ 1 & -1 & -1 & 1 \end{bmatrix} \right)$$

$$= \begin{bmatrix} 0 & 1 & 0 & 0 \end{bmatrix} = \langle 01|$$

Alternatively, the output response in spectral form may be separated into outer product factors and then each transformed with the $\frac{1}{2}\mathbf{H}_1$ inverse transformation matrix.

$$\begin{bmatrix} 1 & -1 & 1 & -1 \end{bmatrix} (\frac{1}{4}\mathbf{H}_2) = \begin{bmatrix} 1 & 1 \end{bmatrix} (\frac{1}{2}\mathbf{H}_1) \otimes \begin{bmatrix} 1 & -1 \end{bmatrix} (\frac{1}{2}\mathbf{H}_1)$$
$$= \begin{bmatrix} 1 & 0 \end{bmatrix} \otimes \begin{bmatrix} 0 & 1 \end{bmatrix} = \langle 0| \otimes \langle 1| = \langle 01|$$

□

If it is desired to compute the spectral coefficients for each individual primary output of a switching network, the corresponding spectral response matrix must first be formed. This can be accomplished be extracting a cone partition from the netlist using the desired primary output as the apex of the cone as shown in Figure 6.6. The procedure for extracting the transfer matrix as previously described is then applied and the resulting transfer matrix characterizes the particular primary output of interest only. Example 6.11 illustrates this process.

Example 6.11   *Walsh Coefficients for Each Network Output*
The spectral coefficient corresponding to netlist output f_1 is computed using the spectral output response matrix for f_1 only denoted as $\mathbf{T}_{sf_1}$. $\mathbf{T}_{sf_1}$ is obtained by extracting the transfer matrix

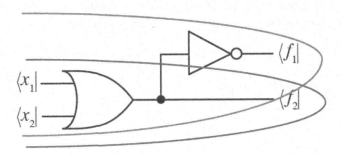

Figure 6.6: Example logic network with primary output cones.

from the portion of the netlist supporting primary output f_1 as shown in the upper blue cone of Figure 6.6. The transfer matrix is extracted in factored form and the calculations are carried out to determine the monolithic explicit form.

$$
\mathbf{T}_{f_1} = (\mathbf{O})(\mathbf{NI}) = \begin{bmatrix} 1 & 0 \\ 0 & 1 \\ 0 & 1 \\ 0 & 1 \end{bmatrix} \begin{bmatrix} 0 & 1 \\ 1 & 0 \end{bmatrix} = \begin{bmatrix} 0 & 1 \\ 1 & 0 \\ 1 & 0 \\ 1 & 0 \end{bmatrix}
$$

The spectral response matrix for netlist output f_2, denoted as $\mathbf{T}_{sf_1}$ is then computed according to the definition.

$$
\mathbf{T}_{sf_1} = \mathbf{TH}_1 = \begin{bmatrix} 0 & 1 \\ 1 & 0 \\ 1 & 0 \\ 1 & 0 \end{bmatrix} \begin{bmatrix} 1 & 1 \\ 1 & -1 \end{bmatrix} = \begin{bmatrix} 1 & -1 \\ 1 & 1 \\ 1 & 1 \\ 1 & 1 \end{bmatrix}
$$

The spectral coefficient $\langle s_{10f_1}|$ can then be computed directly as shown.

$$
\langle s_{10f_1}| = \langle 10|\mathbf{H}_2\mathbf{T}_{sf_1} = \begin{bmatrix} 0 & 0 & 1 & 0 \end{bmatrix} \begin{bmatrix} 1 & 1 & 1 & 1 \\ 1 & -1 & 1 & -1 \\ 1 & 1 & -1 & -1 \\ 1 & -1 & -1 & 1 \end{bmatrix} \begin{bmatrix} 1 & -1 \\ 1 & 1 \\ 1 & 1 \\ 1 & 1 \end{bmatrix}
$$

$$
= \begin{bmatrix} 1 & 1 & -1 & -1 \end{bmatrix} \begin{bmatrix} 1 & -1 \\ 1 & 1 \\ 1 & 1 \\ 1 & 1 \end{bmatrix} = \begin{bmatrix} 0 & -2 \end{bmatrix}
$$

Likewise, the spectral coefficient $\langle s_{10f_2}|$ corresponding to netlist output f_2 is computed using the spectral output response matrix for f_2 only denoted as $\mathbf{T}_{sf_2}$.

$$\langle s_{10f_2}| = \langle 10|\mathbf{H}_2\mathbf{T}_{sf_2} = \begin{bmatrix} 0 & 0 & 1 & 0 \end{bmatrix} \begin{bmatrix} 1 & 1 & 1 & 1 \\ 1 & -1 & 1 & -1 \\ 1 & 1 & -1 & -1 \\ 1 & -1 & -1 & 1 \end{bmatrix} \begin{bmatrix} 1 & 1 \\ 1 & -1 \\ 1 & -1 \\ 1 & -1 \end{bmatrix}$$

$$= \begin{bmatrix} 1 & 1 & -1 & -1 \end{bmatrix} \begin{bmatrix} 1 & 1 \\ 1 & -1 \\ 1 & -1 \\ 1 & -1 \end{bmatrix} = \begin{bmatrix} 0 & 2 \end{bmatrix}$$

□

6.4.4   COMPUTING WALSH SPECTRAL COEFFICIENTS FROM A NETLIST

The concept of a spectral output response matrix and its use in calculating the Walsh spectral coefficients and the network output response were developed using monolithic explicit matrices in the previous section. However, the method of performing the calculation with the spectral response matrix in distributed factored form is possible and provides a useful technique for determining spectral coefficients using the vector space model that has not been shown to be possible with the traditional switching algebra models. The significance of the preceding results is that a method for computing a single spectral coefficient with a simple traversal of an unaltered netlist is available. There is no required extraction of a switching function representation, thus all the problems with intermediate data structures for representation of switching functions and spectra are avoided. These problems were the source of one of the most prohibiting factors in utilizing spectral methods in modern EDA tasks.

The process of computing spectral coefficients through netlist traversals is illustrated using the example netlist in Figure 6.5. When the netlist transfer matrix is given symbolically in distributed factored form, it may be graphically represented as shown in Figure 6.7a. Likewise, the spectral response matrix may be represented in distributed factored form by the diagram in Figure 6.7b. The spectral response matrix is defined as shown in Figure 6.7b to provide a matrix that can be used to calculate both the output response due to a stimulus expressed as a canonical basis vector, or to compute a spectral coefficient by first transforming the input stimulus vector to the Walsh domain. However, if it is desired to compute the spectral coefficients only, the distributed form in Figure 6.7c may be more convenient since it allows the input stimulus to be represented as a canonical basis vector and the corresponding output response is then the spectral response.

As an example of computing a spectral coefficient through a traversal of the netlist, we will utilize the form in Figure 6.7c to compute the spectral coefficients $\langle s_{10f_1}|$ and $\langle s_{10f_2}|$. These

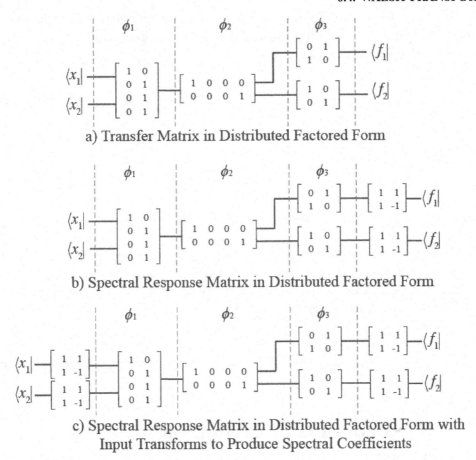

a) Transfer Matrix in Distributed Factored Form

b) Spectral Response Matrix in Distributed Factored Form

c) Spectral Response Matrix in Distributed Factored Form with Input Transforms to Produce Spectral Coefficients

Figure 6.7: Distributed factored form a) transfer matrix b) Walsh spectral response matrix.

are the same spectral coefficients that were computed using the explicit monolithic forms of the matrices in Example 6.11.

Example 6.12 *Walsh Coefficient Computation through Netlist Traversal*
To compute spectral coefficients through the traversal of a netlist, each netlist gate is described with its corresponding transfer matrix. Each interconnecting line is then annotated with values proceeding from the primary inputs to the primary outputs. Initially, a value is assigned to the primary inputs that correspond to the desired spectral coefficient. Since we are interested in the spectral coefficients $\langle s_{10f_1}|$ and $\langle s_{10f_2}|$, we make initial assignments of $\langle x_1| = \langle 1|$ and $\langle x_2| = \langle 0|$. These values are propagated toward the primary outputs and they are appropriately transformed as operator transfer matrices are encountered. When a matrix is encountered that has two or more

input lines, the values on each line are combined using the outer product operation, then a direct vector-matrix product is computed. Figure 6.8 contains a diagram of the graphical distributed factored form with each line annotated.

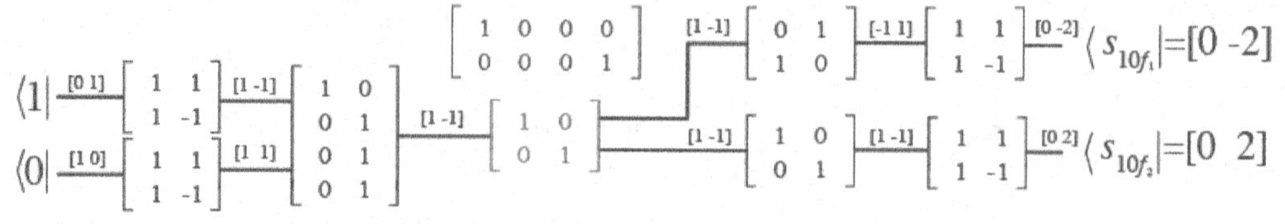

Figure 6.8: Spectral coefficient computed through netlist traversal.

One important modification to note is that the internal fanout transfer matrix $\mathbf{FO}$ is replaced with a pass-through identity matrix $\mathbf{I}$. This is necessary when propagating values through the netlist to discard the null values that would otherwise result. The original $\mathbf{FO}$ matrix is shown in blue above the newly added identity matrix $\mathbf{I}$ shown in red. When the monolithic transfer matrices or partition transfer matrices are being computed, it is necessary to leave the original $\mathbf{FO}$ matrix in place to account for proper vector space dimensioning; however, it is more convenient to model the fanout as a pass-through with two exiting lines when propagating values through a netlist.   □

6.5   THE REED-MULLER TRANSFORM

The Reed-Muller (RM) transform is another popular spectral transform for binary-valued switching networks. The RM transform results in coefficients for a specific representation of a switching function where certain forms of product terms are combined with the Exclusive-OR (XOR) operator. Each RM representation of a switching function contains at most 2^n product terms consisting of all possible combinations of each switching function variable in a specific polarity. The polarity assignment for the set of switching variables $\{x_0, x_1, \ldots, x_{n-1}\}$ defines which of the specific 2^n RM transformations is being applied.

The general form of the Reed-Muller expansion for a switching function of 3-variables is given in Equation 6.3 as

$$f(x_1, x_2, x_3) = r_0(1) \oplus r_1(\dot{x}_1) \oplus r_2(\dot{x}_2) \oplus r_3(\dot{x}_3) \oplus r_{12}(\dot{x}_1\dot{x}_2) \oplus r_{13}(\dot{x}_1\dot{x}_3)$$
$$\oplus r_{23}(\dot{x}_2\dot{x}_3) \oplus r_{123}(\dot{x}_1\dot{x}_2\dot{x}_3) \tag{6.3}$$

where $\dot{x}_i$ indicates that each occurrence of x_i is either in positive polarity form (uncomplemented) or negative polarity form ($\bar{x}_i$) but not both. Considering all possible combinations of polarity for functions of n variables, there are a total of 2^n different polarity RM transformation matrices. The coefficients r_i are the RM spectral transform coefficients and are constants, $r_i \in \mathbb{B} = \{0, 1\}$.

The positive polarity RM transform matrix for functions of one variable can be used as a Kernel to generate higher-ordered transforms in the same way as given in the Walsh transform. Equation 6.4 contains the $\mathbf{R}_1$ transform matrix and the outer product expansion for higher-ordered transformation matrices.

$$\mathbf{R}_1 = \begin{bmatrix} 1 & 0 \\ 1 & 1 \end{bmatrix}, \quad \mathbf{R}_n = \overset{2^n}{\underset{n=1}{\bigotimes}} \mathbf{R}_1 \tag{6.4}$$

The negative polarity transformation matrix can be similarly defined for a function of one variable using

$$\mathbf{R}_{n1} = \begin{bmatrix} 1 & 1 \\ 0 & 1 \end{bmatrix}.$$

6.5.1   RM TRANSFORM OF SCALAR-VALUED SWITCHING FUNCTIONS

Mixed polarity transform matrices can be calculated using $\mathbf{R}_1$ for the positive polarity variables and $\mathbf{R}_{n1}$ for the negative polarity variables. The polarity number indicates whether $\mathbf{R}_1$ or $\mathbf{R}_{n1}$ should be used in the expansion. For a switching function of $n = 3$ variables, the polarity-N transform would be expressed as an outer product of three $\mathbf{R}_1$ or $\mathbf{R}_{n1}$ matrices. To determine which of the matrices is used, the polarity number N is expressed in binary form and the occurrence of a "0" digit indicates the use of $\mathbf{R}_1$ whereas the occurrence of a "1" digit indicates the use of $\mathbf{R}_{n1}$.

Example 6.13   *RM Transform Matrix Example*
The monolithic polarity-6 RM transformation matrix is calculated by expanding 6_{10} into its binary equivalent 110_2. The occurrence of bit value "1" in the first two positions of 110_2 indicates that $\mathbf{R}_{n1}$ is used as the first two outer product factors and $\mathbf{R}_1$ is the last factor.

$$\mathbf{R}_{n1} \otimes \mathbf{R}_{n1} \otimes \mathbf{R}_1 = \begin{bmatrix} 1 & 1 \\ 0 & 1 \end{bmatrix} \otimes \begin{bmatrix} 1 & 1 \\ 0 & 1 \end{bmatrix} \otimes \begin{bmatrix} 1 & 0 \\ 1 & 1 \end{bmatrix}$$

$$= \begin{bmatrix} 1 & 1 \\ 0 & 1 \end{bmatrix} \otimes \begin{bmatrix} 1 & 0 & 1 & 0 \\ 1 & 1 & 1 & 1 \\ 0 & 0 & 1 & 0 \\ 0 & 0 & 1 & 1 \end{bmatrix}$$

$$
= \begin{bmatrix}
1 & 0 & 1 & 0 & 1 & 0 & 1 & 0 \\
1 & 1 & 1 & 1 & 1 & 1 & 1 & 1 \\
0 & 0 & 1 & 0 & 0 & 0 & 1 & 0 \\
0 & 0 & 1 & 1 & 0 & 0 & 1 & 1 \\
0 & 0 & 0 & 0 & 1 & 0 & 1 & 0 \\
0 & 0 & 0 & 0 & 1 & 1 & 1 & 1 \\
0 & 0 & 0 & 0 & 0 & 0 & 1 & 0 \\
0 & 0 & 0 & 0 & 0 & 0 & 1 & 1
\end{bmatrix}
$$

□

One of the most common uses of the RM transform is to determine the spectral coefficients r_i so that a switching function can be expressed in the form of an RM expansion.

Example 6.14   *RM Expansion Example for a Switching Function*
Consider the example switching function characterized by the truth table:

x_2	x_1	x_0	f
0	0	0	1
0	0	1	0
0	1	0	1
0	1	1	0
1	0	0	1
1	0	1	1
1	1	0	0
1	1	1	1

The polarity-6 RM spectrum of this example function is computed using the RM transformation matrix calculated in Example 6.7 and the switching algebra model for the function as:

$$
\begin{bmatrix}
1 & 0 & 1 & 0 & 1 & 0 & 1 & 0 \\
1 & 1 & 1 & 1 & 1 & 1 & 1 & 1 \\
0 & 0 & 1 & 0 & 0 & 0 & 1 & 0 \\
0 & 0 & 1 & 1 & 0 & 0 & 1 & 1 \\
0 & 0 & 0 & 0 & 1 & 0 & 1 & 0 \\
0 & 0 & 0 & 0 & 1 & 1 & 1 & 1 \\
0 & 0 & 0 & 0 & 0 & 0 & 1 & 0 \\
0 & 0 & 0 & 0 & 0 & 0 & 1 & 1
\end{bmatrix}
\begin{bmatrix}
1 \\ 0 \\ 1 \\ 0 \\ 1 \\ 1 \\ 0 \\ 1
\end{bmatrix}
=
\begin{bmatrix}
1 \\ 1 \\ 1 \\ 0 \\ 1 \\ 1 \\ 0 \\ 1
\end{bmatrix}
$$

□

Example 6.14 illustrates the computation of the polarity-6 RM spectrum using the conventional switching algebra model for an example function. Unlike the Walsh spectra, the addition operation is performed modulo-r, or in this case, modulo-2 which is also the familiar exclusive-OR (XOR) operation. The resulting spectrum can then be interpreted as a set of weights, r_i corresponding to the general form of the Reed-Muller expansion given in Equation 6.3. Because the addition operation is modular, all $r_i \in \mathbb{B}$. Each column vector in the transformation matrix corresponds to a particular product term in Equation 6.3 and is the output response of the product terms. From this interpretation, the RM spectrum provides the weights corresponding to each product term. In Example 6.14, six r_i spectral coefficients are non-zero, $r_{\bar{x}_2\bar{x}_1}$, $r_{\bar{x}_2\bar{x}_1x_0}$, $r_{\bar{x}_2}$, $r_{\bar{x}_1}$, $r_{\bar{x}_1x_0}$, and r_{x_0}. Thus, the polarity-6 RM expansion of the example switching function is

$$f(x_2, x_1, x_0) = \bar{x}_2\bar{x}_1 \oplus \bar{x}_2\bar{x}_1x_0 \oplus \bar{x}_2 \oplus \bar{x}_1 \oplus \bar{x}_1x_0 \oplus x_0.$$

Because the RM transform is a discrete Fourier transform over $GF(2)$, a butterfly diagram can be formulated similar to that for the Walsh transform. Additions at the butterfly diagram vertices are performed modulo-2 and each edge multiplier is $\in \mathbb{B}$. For this reason, the butterfly diagram is generally represented with the lack of an edge for the edges with zero weight and with a solid edge for unity-valued weights. Figure 6.9 contains butterfly diagrams for both $\mathbf{R}_1$ and $\mathbf{R}_{n1}$.

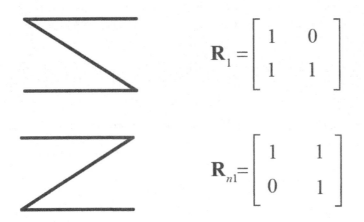

$$\mathbf{R}_1 = \begin{bmatrix} 1 & 0 \\ 1 & 1 \end{bmatrix}$$

$$\mathbf{R}_{n1} = \begin{bmatrix} 1 & 1 \\ 0 & 1 \end{bmatrix}$$

Figure 6.9: $\mathbf{R}_1$, $\mathbf{R}_{n1}$, and corresponding butterfly diagrams.

Figure 6.10 contains the butterfly diagram for the polarity-6 RM transform of a switching function of three variables and is annotated with values from Example 6.14.

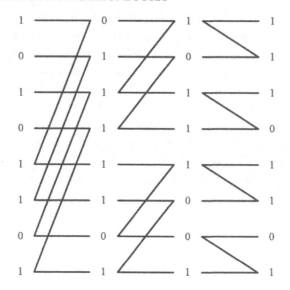

Figure 6.10: RM butterfly diagram for Example 6.14 calculation.

6.5.2   RM TRANSFORM FOR VECTOR-VALUED SWITCHING FUNCTIONS

Vector space models for switching networks may also be transformed to the RM domain. The constants $\langle 0|$ and $\langle 1|$ are transformed into the positive polarity RM spectral domain as follows.

$$\langle 0|\mathbf{R}_1 = \begin{bmatrix} 1 & 0 \end{bmatrix}\begin{bmatrix} 1 & 0 \\ 1 & 1 \end{bmatrix} = \begin{bmatrix} 1 & 0 \end{bmatrix} \quad \langle 1|\mathbf{R}_1 = \begin{bmatrix} 0 & 1 \end{bmatrix}\begin{bmatrix} 1 & 0 \\ 1 & 1 \end{bmatrix} = \begin{bmatrix} 1 & 1 \end{bmatrix}$$

Likewise, the constants $\langle \varnothing|$ and $\langle t|$ may be transformed to the negative-polarity RM domain.

$$\langle \varnothing|\mathbf{R}_1 = \begin{bmatrix} 0 & 0 \end{bmatrix}\begin{bmatrix} 1 & 0 \\ 1 & 1 \end{bmatrix} = \begin{bmatrix} 0 & 0 \end{bmatrix} \quad \langle t|\mathbf{R}_1 = \begin{bmatrix} 1 & 1 \end{bmatrix}\begin{bmatrix} 1 & 0 \\ 1 & 1 \end{bmatrix} = \begin{bmatrix} 0 & 1 \end{bmatrix}$$

The constants may similarly be transformed to the negative-polarity RM domain.

$$\langle 0|\mathbf{R}_1 = \begin{bmatrix} 1 & 0 \end{bmatrix}\begin{bmatrix} 1 & 1 \\ 0 & 1 \end{bmatrix} = \begin{bmatrix} 1 & 1 \end{bmatrix} \quad \langle 1|\mathbf{R}_1 = \begin{bmatrix} 0 & 1 \end{bmatrix}\begin{bmatrix} 1 & 1 \\ 0 & 1 \end{bmatrix} = \begin{bmatrix} 0 & 1 \end{bmatrix}$$

$$\langle \varnothing|\mathbf{R}_1 = \begin{bmatrix} 0 & 0 \end{bmatrix}\begin{bmatrix} 1 & 1 \\ 0 & 1 \end{bmatrix} = \begin{bmatrix} 0 & 0 \end{bmatrix} \quad \langle t|\mathbf{R}_1 = \begin{bmatrix} 1 & 1 \end{bmatrix}\begin{bmatrix} 1 & 1 \\ 0 & 1 \end{bmatrix} = \begin{bmatrix} 1 & 0 \end{bmatrix}$$

The preceding calculations allow Hasse diagrams to be constructed for both the positive- and negative-polarity RM transforms of the constants as depicted in Figure 6.11.

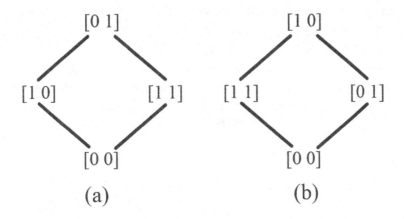

$$(a) \qquad\qquad (b)$$

Figure 6.11: Hasse diagrams of constants in (a) positive- and (b) negative-polarity RM domain.

The Hasse diagrams illustrate an interesting way to define the RM transformation matrix. Note that in the switching domain, input stimuli are modeled as canonical row basis vectors and the collection of all input stimuli when written as a column vector form the identity matrix. For a function of one variable, the collection of all input stimuli are

$$\left[\begin{array}{c} \langle 0| \\ \langle 1| \end{array} \right] = \left[\begin{array}{cc} 1 & 0 \\ 0 & 1 \end{array} \right] = \mathbf{I}.$$

When a matrix is formed of all input stimuli row vectors that are transformed to the polarity-0 RM spectral domain, the polarity-0 RM spectral transformation matrix results.

$$\left[\begin{array}{c} \langle 0|\mathbf{R}_1 \\ \langle 1|\mathbf{R}_1 \end{array} \right] = \left[\begin{array}{cc} 1 & 0 \\ 1 & 1 \end{array} \right] = \mathbf{R}_1$$

Thus, the Hasse diagram implicitly defines the RM spectral transformation matrix by forming a matrix whose row vectors correspond to the middle elements of the Hasse diagram. The middle elements of the Hasse diagram correspond to the distinct singular values (non-null and not the total vector). This same phenomena may be observed when the Hasse diagrams for the Walsh transformed constants are used to specify the Walsh spectral transformation matrix described in the previous section of this chapter.

Computation of the Complete RM Spectrum using the Vector Space Model

The RM spectrum of a switching network modeled in the vector space is a matrix that can be viewed in the form of a column vector whose elements are RM spectral row vectors. Because the transfer matrix is isomorphic to the truth table, the RM spectral response matrix can be defined. The following theorem relates the RM spectrum to the transfer matrix of a logic network.

Theorem 6.15   RM Spectrum Relation to Transfer Matrix
The RM spectrum of a logic network modeled in the vector space is denoted as $\mathbf{S}_f$ and is calculated as the direct matrix product of the RM spectral transformation matrix and the network transfer matrix as given in Equation 6.5 where $\mathbf{R}_n$ represents the RM spectral transformation matrix of n variables for any given polarity and $\mathbf{T}_f$ represents the transfer matrix of a logic network.

$$\mathbf{S}_f = \mathbf{R}_n \mathbf{T}_f \tag{6.5}$$

Proof. The definition of the RM spectrum of a function is the vector produced by the direct vector-matrix product of the RM spectral transformation matrix $\mathbf{R}_n$ and the complete output response vector of a logic network characterized by a vector $|F\rangle$. Stated mathematically,

$$\mathbf{S}_f = \mathbf{R}_n |F\rangle.$$

A single output response of a logic network characterized by $\mathbf{T}_f$ due to the ith input stimulus $\langle x_i|$ is given by $\langle f_i| = \langle x_i|\mathbf{T}_f$. A distinct input stimulus $\langle x_i|$ is modeled as the ith canonic row basis vector and the complete output response $|F\rangle$ is due to all possible canonic basis vectors or input stimuli. A column vector composed of all possible and distinct input stimuli $\langle x_i|$ forms a matrix that is denoted $\mathbf{X}$ where each row vector of $\mathbf{X}$ is a canonic basis vector, hence $\mathbf{X} = \mathbf{I}$. The complete output response $|F\rangle$ is a column vector whose components are row vectors $\langle f_i|$ where each $\langle f_i|$ is the output response due to input stimulus $\langle x_i|$, thus the complete output response is a matrix $\mathbf{F} = |F\rangle$. The complete output response is then given by

$$\mathbf{F} = \mathbf{X}\mathbf{T}_f = \mathbf{I}\mathbf{T}_f = \mathbf{T}.$$

Therefore, the complete output response matrix $\mathbf{F}$ is the transfer matrix $\mathbf{T}_f$. Using this observation with the definition for the calculation of the RM spectrum,

$$\mathbf{S}_f = \mathbf{R}_n \mathbf{F} = \mathbf{R}_n \mathbf{T}_f.$$

□

To illustrate the application of Theorem 6.15, we calculate the polarity-0 RM spectrum of a single output logic network modeled in the vector space in Example 6.16.

Example 6.16   *Polarity-0 RM Spectrum of Logic Network*
Consider the example switching function characterized by the following truth table whose fourth column represents the complete output response using the switching algebraic model and whose fifth column represents the complete output response using the vector space model.

x_2	x_1	x_0	f	$\langle f \vert$
0	0	0	1	$\begin{bmatrix} 0 & 1 \end{bmatrix}$
0	0	1	0	$\begin{bmatrix} 1 & 0 \end{bmatrix}$
0	1	0	1	$\begin{bmatrix} 0 & 1 \end{bmatrix}$
0	1	1	0	$\begin{bmatrix} 1 & 0 \end{bmatrix}$
1	0	0	1	$\begin{bmatrix} 0 & 1 \end{bmatrix}$
1	0	1	1	$\begin{bmatrix} 0 & 1 \end{bmatrix}$
1	1	0	0	$\begin{bmatrix} 1 & 0 \end{bmatrix}$
1	1	1	1	$\begin{bmatrix} 0 & 1 \end{bmatrix}$

By the property of truth table isomorphism, the transfer matrix $\mathbf{T}_f$ is formed as a column vector whose components are row vectors corresponding to the fifth column of the truth table.

$$\mathbf{T}_f = \begin{bmatrix} 0 & 1 \\ 1 & 0 \\ 0 & 1 \\ 1 & 0 \\ 0 & 1 \\ 0 & 1 \\ 1 & 0 \\ 0 & 1 \end{bmatrix}$$

The polarity-0 RM transformation matrix is computed as

$$\mathbf{R}_3 = \begin{bmatrix} 1 & 0 \\ 1 & 1 \end{bmatrix} \otimes \begin{bmatrix} 1 & 0 \\ 1 & 1 \end{bmatrix} \otimes \begin{bmatrix} 1 & 0 \\ 1 & 1 \end{bmatrix} = \begin{bmatrix} 1 & 0 & 0 & 0 & 0 & 0 & 0 & 0 \\ 1 & 1 & 0 & 0 & 0 & 0 & 0 & 0 \\ 1 & 0 & 1 & 0 & 0 & 0 & 0 & 0 \\ 1 & 1 & 1 & 1 & 0 & 0 & 0 & 0 \\ 1 & 0 & 0 & 0 & 1 & 0 & 0 & 0 \\ 1 & 1 & 0 & 0 & 1 & 1 & 0 & 0 \\ 1 & 0 & 1 & 0 & 1 & 0 & 1 & 0 \\ 1 & 1 & 1 & 1 & 1 & 1 & 1 & 1 \end{bmatrix}$$

Using the result of Theorem 6.15 which is given in Equation 6.5, we calculate the polarity-0 RM spectrum of the logic network as follows.

$$
\mathbf{S}_f = \mathbf{R}_3\mathbf{T}_f =
\begin{bmatrix}
1 & 0 & 0 & 0 & 0 & 0 & 0 & 0 \\
1 & 1 & 0 & 0 & 0 & 0 & 0 & 0 \\
1 & 0 & 1 & 0 & 0 & 0 & 0 & 0 \\
1 & 1 & 1 & 1 & 0 & 0 & 0 & 0 \\
1 & 0 & 0 & 0 & 1 & 0 & 0 & 0 \\
1 & 1 & 0 & 0 & 1 & 1 & 0 & 0 \\
1 & 0 & 1 & 0 & 1 & 0 & 1 & 0 \\
1 & 1 & 1 & 1 & 1 & 1 & 1 & 1
\end{bmatrix}
\begin{bmatrix}
0 & 1 \\
1 & 0 \\
0 & 1 \\
1 & 0 \\
0 & 1 \\
0 & 1 \\
1 & 0 \\
0 & 1
\end{bmatrix}
=
\begin{bmatrix}
0 & 1 \\
1 & 1 \\
0 & 0 \\
0 & 0 \\
0 & 0 \\
1 & 1 \\
1 & 1 \\
1 & 1
\end{bmatrix}
$$

The RM spectrum of a switching network modeled in the vector space results in a matrix $\mathbf{S}_f$ whose row vectors are individual RM spectral coefficients while the RM spectrum of a switching network modeled with switching algebra results in scalar-valued RM spectral coefficients with all coefficients $r_i \in \mathbb{B}$. However, there is a relation between the RM spectral coefficients resulting from the two models.

Due to the definition of switching constants, $\langle 0| = \begin{bmatrix} 1 & 0 \end{bmatrix}$ and $\langle 1| = \begin{bmatrix} 0 & 1 \end{bmatrix}$, the second component of each row vector is identical to the scalar constants in the switching algebra models. The result of this definition is that the second column vector in a transfer matrix $\mathbf{T}_f$ is identical to the output response of the switching network when modeled using conventional switching algebra. Additionally, the first column vector in $\mathbf{T}_f$ corresponds to the output response of $\bar{f}$, the inverse or complement of the switching algebraic model of the logic network. These observations are actually corollaries of the property of truth table isomorphism and lead to the following lemma.

Lemma 6.17   RM Spectrum Column Vectors are Related
The RM spectrum of a single-output switching network modeled in the vector space is in the form of a $2^n \times 2$ matrix $\mathbf{S}_f$ whose first column vector is the RM spectrum of the switching algebraic model of the inverse or complement switching function $\bar{f}$ characterizing the network and whose second column vector is the RM spectrum of the switching function f characterizing the network.

Proof. The complete output response of a single-output switching network modeled in the vector space is the transfer matrix $\mathbf{T}_f$ whose second column vector is identical to the switching algebraic model output response and whose first column vector is the inverse of the switching algebraic model. From Theorem 6.15, the RM spectrum is given as $\mathbf{S}_f = \mathbf{R}_n\mathbf{T}_f$, thus the first column vector of $\mathbf{S}_f$ is identical to the scalar RM spectrum of $\bar{f}$ and the second column vector of $\mathbf{S}_f$ is identical to the scalar RM spectrum of f. Where f is the switching function that models the network using conventional Boolean algebra.   □

A further relationship exists between the first and second column vectors of the RM spectrum of a single-output switching network and is stated in Theorem 6.18.

Theorem 6.18   Relation of Column Vectors in RM Spectral Matrix
The column vectors comprising $\mathbf{S}_f$ are of dimension 2^n and exactly $2^n - 1$ components of the first and second column vectors comprising $\mathbf{S}_f$ are identical. An alternative statement of this theorem is that $2^n - 1$ coefficients are of the form $\begin{bmatrix} 1 & 1 \end{bmatrix}$ or $\begin{bmatrix} 0 & 0 \end{bmatrix}$ and the remaining coefficient is of the form $\begin{bmatrix} 0 & 1 \end{bmatrix}$ or $\begin{bmatrix} 1 & 0 \end{bmatrix}$.

Proof. From Lemma 6.17, the column vectors of $\mathbf{S}_f$ are nearly identical with the exception of a single component since the column vectors represent the scalar RM spectra of $\bar{f}$ and f respectively. From the property of the XOR operation, $1 \oplus f = \bar{f}$. The RM expansion given in Equation 6.3 indicates that one of the product terms is the constant 1 function and has a corresponding spectral coefficient r_0. The r_0 coefficient then indicates complementation of the function formed by the collection of the remaining $2^n - 1$ terms in the RM expansion given in Equation 6.3 and it is the r_0 spectral coefficient that differs among the two column vectors comprising $\mathbf{S}_f$. $\square$

The result of Theorem 6.18 can be observed by using the calculation in Example 6.15 to express the switching algebra functions f and $\bar{f}$ that model the switching network. In the case of the polarity-0 RM spectrum, the first element in the column vectors of $\mathbf{S}_f$ corresponds to the r_0 coefficient in the RM expansion. We can write the switching algebra expression for $\bar{f}$ in an exclusive-OR sum-of-products form by including those product terms with RM spectral coefficients of 1. Examination of the calculation in Example 6.15 indicates that $r_3 = r_{13} = r_{12} = r_{123} = 1$; these polarity-0 RM spectral coefficients correspond to product terms x_0, x_2x_0, x_2x_1, and $x_2x_1x_0$ respectively.

$$\bar{f} = x_0 \oplus x_2x_0 \oplus x_2x_1 \oplus x_2x_1x_0$$

Due to the fact that $1 \oplus \bar{f} = f$, complementing the above expression for $\bar{f}$ can be accomplished by simply including a product term of 1. From a spectral point of view, this indicates that $r_0 = 1$ should be included in the second column vector of $\mathbf{S}_f$ which is indeed the case. Thus the RM expansion for f is written as follows.

$$f = 1 \oplus x_0 \oplus x_2x_0 \oplus x_2x_1 \oplus x_2x_1x_0$$

All RM spectra include the r_0 coefficient regardless of polarity value. However, the position of the r_0 spectral coefficient differs among the row vectors comprising $\mathbf{S}_f$ and is dependent upon the particular polarity value. The position of the vector-valued r_0 spectral coefficient within $\mathbf{S}_f$ is identical to the column vector index of the all-1's column vector within the *naturally* ordered

RM transformation matrix. An RM spectral transformation matrix is *naturally* ordered when it is computed as an outer product of R_1 and R_{n1} matrices. The all-1's column vector is that vector whose components are all unity-valued (1) and represents the constant-1 product term in the RM expansion. For example, the polarity-0 RM transformation matrix results in the r_0 coefficient being the first component in the spectrum S_f and has row index 0. We note that this is one of the reasons we utilize the somewhat unconventional manner of indexing rows and columns beginning with zero (0) instead of one (1). It is also noted that the row order for a given RM spectral transformation matrix may be permuted corresponding to a permuted order of the corresponding spectral coefficients when they are calculated using the vector-matrix direct product. When the RM spectral transformation matrix row order is a result of the outer product definition, the so-called *natural* order of the spectral coefficients results. The position of the r_0 spectral coefficient is always present in S_f at the row index equivalent to the column vector index of the all-1's vector within the transformation matrix R_n when R_n is naturally ordered. To demonstrate this observation, Example 6.19 contains the computation of the polarity-6 RM spectrum for the same single-output switching network used in Example 6.16.

Example 6.19   *RM Polarity-6 Spectral Response Matrix*
The example function is characterized by the following truth table:

x_2	x_1	x_0	f	$\langle f \vert$
0	0	0	1	$[\begin{matrix} 0 & 1 \end{matrix}]$
0	0	1	0	$[\begin{matrix} 1 & 0 \end{matrix}]$
0	1	0	1	$[\begin{matrix} 0 & 1 \end{matrix}]$
0	1	1	0	$[\begin{matrix} 1 & 0 \end{matrix}]$
1	0	0	1	$[\begin{matrix} 0 & 1 \end{matrix}]$
1	0	1	1	$[\begin{matrix} 0 & 1 \end{matrix}]$
1	1	0	0	$[\begin{matrix} 1 & 0 \end{matrix}]$
1	1	1	1	$[\begin{matrix} 0 & 1 \end{matrix}]$

The polarity-6 spectral response matrix is calculated as

$$S_f = R_3 T_f = \begin{bmatrix} 1 & 0 & 1 & 0 & 1 & 0 & 1 & 0 \\ 1 & 1 & 1 & 1 & 1 & 1 & 1 & 1 \\ 0 & 0 & 1 & 0 & 0 & 0 & 1 & 0 \\ 0 & 0 & 1 & 1 & 0 & 0 & 1 & 1 \\ 0 & 0 & 0 & 0 & 1 & 0 & 1 & 0 \\ 0 & 0 & 0 & 0 & 1 & 1 & 1 & 1 \\ 0 & 0 & 0 & 0 & 0 & 0 & 1 & 0 \\ 0 & 0 & 0 & 0 & 0 & 0 & 1 & 1 \end{bmatrix} \begin{bmatrix} 0 & 1 \\ 1 & 0 \\ 0 & 1 \\ 1 & 0 \\ 0 & 1 \\ 0 & 1 \\ 1 & 0 \\ 0 & 1 \end{bmatrix} = \begin{bmatrix} 1 & 1 \\ 1 & 1 \\ 1 & 1 \\ 0 & 0 \\ 1 & 1 \\ 1 & 1 \\ 1 & 0 \\ 1 & 1 \end{bmatrix}$$

Because the example switching network has three primary inputs, the size of the RM spectral transformation matrix is $2^3 \times 2^3$ and the row and column indices run from zero to seven. Likewise, the $2^3 = 8$ RM spectral coefficients are contained within S_f as row vectors $\langle s_i |$ that are likewise indexed with $0 \le i \le 7$. The polarity-6 RM spectral transformation matrix contains the all-1's column vector at column index 6 and the corresponding r_0 spectral coefficient occurs within the calculated spectrum S_f at row index 6. As indicated by Theorem 6.18, the only RM spectral coefficient that is not of the form $\begin{bmatrix} 1 & 1 \end{bmatrix}$ or $\begin{bmatrix} 0 & 0 \end{bmatrix}$ is that corresponding to r_0.   □

An RM spectral transformation calculation can be carried out for multiple-output switching networks through the application of Equation 6.5. The transfer matrix for a switching network with n primary inputs and m primary outputs is of dimension $2^n \times 2^m$. Therefore, the complete RM spectrum S_f also has dimensions $2^n \times 2^m$. Each row vector comprising S_f is a distinct RM spectral coefficient and for the case of a multiple output switching network, is in the form of a 2^m dimensional row vector. Example 6.20 contains the calculation of the polarity-0 RM spectrum for the example multiple-output switching network depicted in Figure 6.5.

Example 6.20   *Multi-output Network RM Spectrum*
The transfer matrix for the network depicted in Figure 6.5 is

$$
T_{f_1 f_2} = \begin{bmatrix} \langle 10| \\ \langle 01| \\ \langle 01| \\ \langle 01| \end{bmatrix} = \begin{bmatrix} \begin{bmatrix} 0 & 1 \end{bmatrix} \otimes \begin{bmatrix} 1 & 0 \end{bmatrix} \\ \begin{bmatrix} 1 & 0 \end{bmatrix} \otimes \begin{bmatrix} 0 & 1 \end{bmatrix} \\ \begin{bmatrix} 1 & 0 \end{bmatrix} \otimes \begin{bmatrix} 0 & 1 \end{bmatrix} \\ \begin{bmatrix} 1 & 0 \end{bmatrix} \otimes \begin{bmatrix} 0 & 1 \end{bmatrix} \end{bmatrix} = \begin{bmatrix} 0 & 0 & 1 & 0 \\ 0 & 1 & 0 & 0 \\ 0 & 1 & 0 & 0 \\ 0 & 1 & 0 & 0 \end{bmatrix}.
$$

Equation 6.5 may be used with the polarity-0 RM transformation matrix resulting in the spectrum S_f.

$$
S_f = R_2 T_{f_1 f_2} = \begin{bmatrix} 1 & 0 & 0 & 0 \\ 1 & 1 & 0 & 0 \\ 1 & 0 & 1 & 0 \\ 1 & 1 & 1 & 1 \end{bmatrix} \begin{bmatrix} 0 & 0 & 1 & 0 \\ 0 & 1 & 0 & 0 \\ 0 & 1 & 0 & 0 \\ 0 & 1 & 0 & 0 \end{bmatrix} = \begin{bmatrix} 0 & 0 & 1 & 0 \\ 0 & 1 & 1 & 0 \\ 0 & 1 & 1 & 0 \\ 0 & 1 & 1 & 0 \end{bmatrix}
$$

To qualitatively understand the meaning of RM coefficients for a multiple-output switching network, we can express the above computation as follows.

$$
S_f = R_2 T_{f_1 f_2} = (R_1 \otimes R_1) \begin{bmatrix} \begin{bmatrix} 0 & 1 \end{bmatrix} \otimes \begin{bmatrix} 1 & 0 \end{bmatrix} \\ \begin{bmatrix} 1 & 0 \end{bmatrix} \otimes \begin{bmatrix} 0 & 1 \end{bmatrix} \\ \begin{bmatrix} 1 & 0 \end{bmatrix} \otimes \begin{bmatrix} 0 & 1 \end{bmatrix} \\ \begin{bmatrix} 1 & 0 \end{bmatrix} \otimes \begin{bmatrix} 0 & 1 \end{bmatrix} \end{bmatrix}
$$

From the above expression, it is apparent that each spectral coefficient is a row vector comprised of components that depend upon both f_1 and f_2. □

If the individual RM spectra are desired for outputs $\langle f_1 |$ and $\langle f_2 |$, cones may be formed to determine the dependencies of the outputs as depicted in Figure 6.6 and the method for partitioning and traversing each cone may be invoked to determine the individual transfer matrices. Once the individual transfer matrices are determined, Equation 6.5 may be used to compute the spectra for each cone or primary output. This same method may be used to compute the spectra for internal lines in the netlist when spectra are desired for internal points in a switching network. The transfer matrices that support each individual output are given in symbolic factored form as $\mathbf{T}_{f_1} = (\mathbf{O})(\mathbf{NI})$ and $\mathbf{T}_{f_2} = \mathbf{O}$. In explicit monolithic form, these transfer matrices are given as

$$\mathbf{T}_{f_1} = \begin{bmatrix} 1 & 0 \\ 0 & 1 \\ 0 & 1 \\ 0 & 1 \end{bmatrix} \begin{bmatrix} 0 & 1 \\ 1 & 0 \end{bmatrix} = \begin{bmatrix} 0 & 1 \\ 1 & 0 \\ 1 & 0 \\ 1 & 0 \end{bmatrix} \qquad \mathbf{T}_{f_2} = \begin{bmatrix} 1 & 0 \\ 0 & 1 \\ 0 & 1 \\ 0 & 1 \end{bmatrix}$$

The polarity-0 RM spectra for primary outputs $\langle f_1 |$ and $\langle f_2 |$ are denoted as $\mathbf{S}_{f_1}$ and $\mathbf{S}_{f_2}$ and are calculated using Equation 6.5 as follows.

$$\mathbf{S}_{f_1} = \mathbf{R}_2 \mathbf{T}_{f_1} = \begin{bmatrix} 1 & 0 & 0 & 0 \\ 1 & 1 & 0 & 0 \\ 1 & 0 & 1 & 0 \\ 1 & 1 & 1 & 1 \end{bmatrix} \begin{bmatrix} 0 & 1 \\ 1 & 0 \\ 1 & 0 \\ 1 & 0 \end{bmatrix} = \begin{bmatrix} 0 & 1 \\ 1 & 1 \\ 1 & 1 \\ 1 & 1 \end{bmatrix}$$

$$\mathbf{S}_{f_2} = \mathbf{R}_2 \mathbf{T}_{f_2} = \begin{bmatrix} 1 & 0 & 0 & 0 \\ 1 & 1 & 0 & 0 \\ 1 & 0 & 1 & 0 \\ 1 & 1 & 1 & 1 \end{bmatrix} \begin{bmatrix} 1 & 0 \\ 0 & 1 \\ 0 & 1 \\ 0 & 1 \end{bmatrix} = \begin{bmatrix} 1 & 0 \\ 1 & 1 \\ 1 & 1 \\ 1 & 1 \end{bmatrix}$$

6.5.3 REED-MULLER SPECTRAL RESPONSE MATRICES

The RM spectra can be calculated by first extracting the transfer matrix, then constructing the RM spectral transformation matrix for the desired polarity, and finally applying Equation 6.5. A similar process, but alternative viewpoint, is to construct a transfer matrix that characterizes a switching network in the RM spectral domain in a manner analogous to the switching transfer matrix that characterizes the network in the switching domain. The matrix that characterizes a switching network in the RM domain is referred to as the *Reed–Muller Spectral Response Matrix*. The RM spectral response matrix allows for a network excitation to be modeled as vector and when multiplied with the RM spectral response matrix yields the network response in the RM spectral domain.

Definition 6.21 *RM Spectral Response Matrix*
The RM spectral response matrix of a switching network is identical to the RM spectrum of the network and is given as $S_f = R_n T_f$ where T is the network transfer matrix and R_n is an RM spectral transformation matrix of some desired polarity. □

The above definition indicates that the spectral response matrix is identical to the RM spectrum S_f. While the calculation of the spectral response matrix is identical to that of computing the complete spectrum, S_f represents a different viewpoint in that it is a form of transfer matrix that yields network response values in the RM spectral domain when excited with some specified input stimulus. The input stimuli may be specified in either the switching domain or the RM spectral domain as shown in the following example.

Example 6.22 *RM Spectral Response Example for Single Output Network*
In this example, we consider a switching network with three primary inputs and single primary output modeled in the vector space and characterized by the following table.

$\langle x_2 \vert$	$\langle x_1 \vert$	$\langle x_0 \vert$	$\langle f \vert$	$\langle f \vert$
$\langle 0 \vert$	$\langle 0 \vert$	$\langle 0 \vert$	$\langle 1 \vert$	$\begin{bmatrix} 0 & 1 \end{bmatrix}$
$\langle 0 \vert$	$\langle 0 \vert$	$\langle 1 \vert$	$\langle 0 \vert$	$\begin{bmatrix} 1 & 0 \end{bmatrix}$
$\langle 0 \vert$	$\langle 1 \vert$	$\langle 0 \vert$	$\langle 1 \vert$	$\begin{bmatrix} 0 & 1 \end{bmatrix}$
$\langle 0 \vert$	$\langle 1 \vert$	$\langle 1 \vert$	$\langle 0 \vert$	$\begin{bmatrix} 1 & 0 \end{bmatrix}$
$\langle 1 \vert$	$\langle 0 \vert$	$\langle 0 \vert$	$\langle 1 \vert$	$\begin{bmatrix} 0 & 1 \end{bmatrix}$
$\langle 1 \vert$	$\langle 0 \vert$	$\langle 1 \vert$	$\langle 1 \vert$	$\begin{bmatrix} 0 & 1 \end{bmatrix}$
$\langle 1 \vert$	$\langle 1 \vert$	$\langle 0 \vert$	$\langle 0 \vert$	$\begin{bmatrix} 1 & 0 \end{bmatrix}$
$\langle 1 \vert$	$\langle 1 \vert$	$\langle 1 \vert$	$\langle 1 \vert$	$\begin{bmatrix} 0 & 1 \end{bmatrix}$

The polarity-0 spectral response matrix is calculated as

$$S_f = R_3 T_f = \begin{bmatrix} 1 & 0 & 0 & 0 & 0 & 0 & 0 & 0 \\ 1 & 1 & 0 & 0 & 0 & 0 & 0 & 0 \\ 1 & 0 & 1 & 0 & 0 & 0 & 0 & 0 \\ 1 & 1 & 1 & 1 & 0 & 0 & 0 & 0 \\ 1 & 0 & 0 & 0 & 1 & 0 & 0 & 0 \\ 1 & 1 & 0 & 0 & 1 & 1 & 0 & 0 \\ 1 & 0 & 1 & 0 & 1 & 0 & 1 & 0 \\ 1 & 1 & 1 & 1 & 1 & 1 & 1 & 1 \end{bmatrix} \begin{bmatrix} 0 & 1 \\ 1 & 0 \\ 0 & 1 \\ 1 & 0 \\ 0 & 1 \\ 0 & 1 \\ 1 & 0 \\ 0 & 1 \end{bmatrix} = \begin{bmatrix} 0 & 1 \\ 1 & 1 \\ 0 & 0 \\ 0 & 0 \\ 0 & 0 \\ 1 & 1 \\ 1 & 1 \\ 1 & 1 \end{bmatrix}.$$

Each row vector in $\mathbf{S}_f$ is a polarity-0 RM spectral coefficient in the vector space model. Using the result of Lemma 6.17 that describes the relationship between the RM spectral coefficients in the vector space with those in the switching algebraic model, the rightmost column vector of $\mathbf{S}_f$ can be used to express the functionality of the network as a polarity-0 RM expansion with the conventional switching algebra model according to Equation 6.3 as

$$f(x_2, x_1, x_0) = 1 \oplus x_0 \oplus x_2x_0 \oplus x_2x_1 \oplus x_2x_1x_0.$$

Likewise, the leftmost column vector of $\mathbf{S}_f$ can also be used according to Lemma 6.17 to obtain

$$\bar{f}(x_2, x_1, x_0) = x_0 \oplus x_2x_0 \oplus x_2x_1 \oplus x_2x_1x_0.$$

However, from the point of view of using $\mathbf{S}_f$ as a polarity-0 RM spectral response matrix, we use the matrix to characterize the transfer properties of the network in the RM spectral domain when excited with a stimulus input vector. When the input vector is expressed as a switching input stimulus for the case $\langle x_2x_1x_0| = \langle 111| = \begin{bmatrix} 0 & 0 & 0 & 0 & 0 & 0 & 0 & 1 \end{bmatrix}$, we can obtain the vector-valued spectral coefficient $\langle r_{123}|$.

$$\langle r_{123}| = \langle x_2x_1x_0|\mathbf{S}_f = \begin{bmatrix} 0 & 0 & 0 & 0 & 0 & 0 & 0 & 1 \end{bmatrix} \begin{bmatrix} 0 & 1 \\ 1 & 1 \\ 0 & 0 \\ 0 & 0 \\ 0 & 0 \\ 1 & 1 \\ 1 & 1 \\ 1 & 1 \end{bmatrix} = \begin{bmatrix} 1 & 1 \end{bmatrix}$$

When the input stimulus is expressed in the RM polarity-0 domain, we can excite the spectral response matrix and compute the output response in the switching domain. The RM polarity-0 spectral representation of the input stimulus is

$$\langle x_2 x_1 x_0 | \mathbf{R}_3 = \begin{bmatrix} 0 & 0 & 0 & 0 & 0 & 0 & 0 & 1 \end{bmatrix} \begin{bmatrix} 1 & 0 & 0 & 0 & 0 & 0 & 0 & 0 \\ 1 & 1 & 0 & 0 & 0 & 0 & 0 & 0 \\ 1 & 0 & 1 & 0 & 0 & 0 & 0 & 0 \\ 1 & 1 & 1 & 1 & 0 & 0 & 0 & 0 \\ 1 & 0 & 0 & 0 & 1 & 0 & 0 & 0 \\ 1 & 1 & 0 & 0 & 1 & 1 & 0 & 0 \\ 1 & 0 & 1 & 0 & 1 & 0 & 1 & 0 \\ 1 & 1 & 1 & 1 & 1 & 1 & 1 & 1 \end{bmatrix}$$

$$= \begin{bmatrix} 1 & 1 & 1 & 1 & 1 & 1 & 1 & 1 \end{bmatrix}.$$

Applying $\langle x_2 x_1 x_0 | \mathbf{R}_3$ as an input stimulus to the spectral response matrix yields

$$\langle r_{123} | = \langle x_2 x_1 x_0 | \mathbf{R}_3 \mathbf{S}_f = \begin{bmatrix} 1 & 1 & 1 & 1 & 1 & 1 & 1 & 1 \end{bmatrix} \begin{bmatrix} 0 & 1 \\ 1 & 1 \\ 0 & 0 \\ 0 & 0 \\ 0 & 0 \\ 1 & 1 \\ 1 & 1 \\ 1 & 1 \end{bmatrix} = \begin{bmatrix} 0 & 1 \end{bmatrix}.$$

Thus, we obtain the output response of the network in the switching domain that could similarly be obtained as $\langle x_2 x_1 x_0 | \mathbf{T}_f$. The importance of this example is to demonstrate that the polarity-0 RM spectral response matrix can be used to compute a switching network output response in either the RM spectral domain or the switching domain depending on whether the input stimulus is specified in the switching or RM spectral domain.   □

The result of example 6.22 can be proven to hold for any given switching network modeled in the vector space as is accomplished with Theorem 6.23.

Theorem 6.23   RM Spectral Response Matrix for Switching Domain Output Response
The output response of a switching network can be calculated by exciting the RM spectral response matrix with an input stimulus expressed in the RM spectral domain.

$$(\langle x_1 x_2 \ldots x_n | \mathbf{R}_n) \mathbf{S}_f = \langle x_2 x_1 x_0 | \mathbf{T}_f = \langle f_1 f_2 \ldots f_m | \tag{6.6}$$

Proof. An input stimulus vector $\langle x_1 x_2 \ldots x_n|$ expressed in the RM spectral domain is calculated as $\langle x_1 x_2 \ldots x_n | \mathbf{R}_n$. The theorem states that this vector, when multiplied with the RM spectral response matrix, $\mathbf{S}_f$, results in the network output response $\langle f_1 f_2 \ldots f_m|$. Applying the definition of the RM spectral response matrix, $\mathbf{S}_f = \mathbf{R}_n \mathbf{T}_f$, to the theorem statement results in

$$\langle x_1 x_2 \ldots x_n | \mathbf{R}_n \mathbf{S}_f = \langle x_1 x_2 \ldots x_n | \mathbf{R}_n \mathbf{R}_n \mathbf{T}_f = \langle f_1 f_2 \ldots f_m|.$$

Because $\mathbf{R}_n$ is orthonormal and symmetric, $\mathbf{R}_n^{-1} = \mathbf{R}_n$. Therefore $\mathbf{R}_n \mathbf{R}_n = \mathbf{I}$. Substituting this result into the previous equation yields

$$\langle x_1 x_2 \ldots x_n | \mathbf{R}_n \mathbf{R}_n \mathbf{T}_f = \langle x_1 x_2 \ldots x_n | \mathbf{I} \mathbf{T}_f = \langle x_1 x_2 \ldots x_n | \mathbf{T}_f = \langle f_1 f_2 \ldots f_m|$$

which is the definition of the transfer matrix that characterizes a switching network. □

Computation of Subsets of RM Spectral Coefficients

Individual or subsets of RM spectral coefficients may be computed using the spectral response matrix $\mathbf{S}_f$ and a vector that indexes a particular coefficient or subset of coefficients. For functions of n variables, there are 2^n distinct RM spectral coefficients that serve as coefficients of the RM expansion of a switching function. Within the vector space model, these are denoted as $\langle r_{d_{k-1} d_{k-2} \ldots d_1 d_0}|$ where the k_i digits correspond to the particular variables that form a product term. For example $\langle r_{13}|$ is the RM coefficient that corresponds to product term $x_1 x_3$.

When the RM spectral response matrix characterizing a switching network is used to compute an RM spectral coefficient, it is multiplied by an indexing row vector that points to the desired coefficient. All RM spectral transformations matrices in this book are constructed with natural ordering due to the outer product definition. The natural ordering for an $n = 3$ polarity-0 spectral response matrix is

$$\begin{bmatrix} \langle r_0| \\ \langle r_3| \\ \langle r_2| \\ \langle r_{23}| \\ \langle r_1| \\ \langle r_{13}| \\ \langle r_{12}| \\ \langle r_{123}| \end{bmatrix}.$$

Based upon the natural ordering of the polarity-0 RM coefficients, the index vector can be formulated as $\langle (b_1 b_2 b_3)_2|$ where the bit string $b_1 b_2 b_3$ contains $b_i = 1$ to indicate the presence of variable x_i in the product term of the RM expansion that corresponds to the selected RM spectral coefficient, and $b_i = 0$ if the corresponding variable is not present in the product term. As an example, the index vector for the polarity-0 RM spectral coefficient r_{12} is $(110)_2$ since this coefficient corresponds to $r_{12}x_1x_2$ in the expansion and x_3 is not present.

As previously discussed, natural ordering results in a permutation of the coefficient ordering according to the polarity number. This permutation is easily computed by computing the bitwise exclusive-OR of $b_1b_2b_3$ with the binary value of the polarity number. For the polarity-0 case, the index vectors are ordered as ascending binary values beginning with $\langle 00\ldots 0|$ and ending with $\langle 11\ldots 1|$, and the bitwise exclusive-OR with the polarity value $(00\ldots 0)_2$ leaves this ordering the same. For a polarity-p index vector, the polarity-0 index is first formed and then permuted by performing a bitwise exclusive-OR operation with the binary value of p.

It is noted that this method of computing the index vector exactly corresponds to the input stimulus vector $\langle x_1x_2\ldots x_n|$ when the spectral response matrix is used to compute an output response rather than to determine a spectral coefficient.

Example 6.24   *Computation of Polarity-0 Coefficient r_{13} for an Example Network*
It is desired to compute the RM polarity-0 spectral coefficient r_{13} for the switching network characterized by the spectral response matrix given in the previous Example 6.22. To perform this computation, the spectral response matrix is formulated using the desired polarity number and it is then multiplied with an index vector that points to coefficient r_{13}. This index vector is calculated as $\langle (101) \oplus (000)| = \langle 101|$.

$$\langle 1| \otimes \langle 0| \otimes \langle 1| = \begin{bmatrix} 0 & 1 \end{bmatrix} \otimes \begin{bmatrix} 1 & 0 \end{bmatrix} \otimes \begin{bmatrix} 0 & 1 \end{bmatrix} = \begin{bmatrix} 0 & 0 & 0 & 0 & 0 & 1 & 0 & 0 \end{bmatrix}$$

$$\langle r_{13}| = \langle 101|\mathbf{S}_f = \begin{bmatrix} 0 & 0 & 0 & 0 & 0 & 1 & 0 & 0 \end{bmatrix} \begin{bmatrix} 0 & 1 \\ 1 & 1 \\ 0 & 0 \\ 0 & 0 \\ 0 & 0 \\ 1 & 1 \\ 1 & 1 \\ 1 & 1 \end{bmatrix} = \begin{bmatrix} 1 & 1 \end{bmatrix}$$

From the point of view of the polarity-0 RM expansion, the result of this computation indicates that the product term x_1x_3 is included in the conventional switching algebraic RM expansion of both $f(x_1x_2x_3)$ and $\bar{f}(x_1x_2x_3)$.   □

Example 6.25   *Computation of Polarity-6 r_{13} for an Example Network*

A similar calculation is performed when spectral coefficients are desired for other RM polarity values. In this example, we consider polarity-6. The polarity-6 RM expansion involves product terms comprised of the variables $\{\bar{x}_1, \bar{x}_2, x_3\}$. The polarity-6 spectral response matrix for the example function used in the previous example 6.24 is computed as

$$
\mathbf{S}_f = \mathbf{R}_3 \mathbf{T}_f =
\begin{bmatrix}
1 & 0 & 1 & 0 & 1 & 0 & 1 & 0 \\
1 & 1 & 1 & 1 & 1 & 1 & 1 & 1 \\
0 & 0 & 1 & 0 & 0 & 0 & 1 & 0 \\
0 & 0 & 1 & 1 & 0 & 0 & 1 & 1 \\
0 & 0 & 0 & 0 & 1 & 0 & 1 & 0 \\
0 & 0 & 0 & 0 & 1 & 1 & 1 & 1 \\
0 & 0 & 0 & 0 & 0 & 0 & 1 & 0 \\
0 & 0 & 0 & 0 & 0 & 0 & 1 & 1
\end{bmatrix}
\begin{bmatrix}
0 & 1 \\
1 & 0 \\
0 & 1 \\
1 & 0 \\
0 & 1 \\
0 & 1 \\
1 & 0 \\
0 & 1
\end{bmatrix}
=
\begin{bmatrix}
1 & 1 \\
1 & 1 \\
1 & 1 \\
0 & 0 \\
1 & 1 \\
1 & 1 \\
1 & 0 \\
1 & 1
\end{bmatrix}.
$$

The naturally ordered polarity-6 RM spectral coefficients comprising the spectral response matrix for the example function are

$$
\begin{bmatrix}
1 & 1 \\
1 & 1 \\
1 & 1 \\
0 & 0 \\
1 & 1 \\
1 & 1 \\
1 & 0 \\
1 & 1
\end{bmatrix}
=
\begin{bmatrix}
\langle r_{12}| \\
\langle r_{123}| \\
\langle r_{1}| \\
\langle r_{13}| \\
\langle r_{2}| \\
\langle r_{23}| \\
\langle r_{0}| \\
\langle r_{3}|
\end{bmatrix}.
$$

It is observed that r_{13} occurs at row index value three, where the row indices run from zero to seven for this $n = 3$ primary input switching network. The polarity-6 index vector for coefficient r_{13} is computed as $\langle (101) \oplus (110)| = \langle 011|$. The 110 bit string corresponds to polarity-6 and the exclusive-OR operation permutes the polarity-0 index vector $\langle 101|$ accordingly. The polarity-6 spectral coefficient may now be calculated from the spectral response matrix as $\langle 011|\mathbf{S}_f$.

$$r_{13} = (((101) \oplus (110)|)\mathbf{S}_f = \langle 011|\mathbf{S}_f$$

$$= \begin{bmatrix} 0 & 0 & 0 & 1 & 0 & 0 & 0 & 0 \end{bmatrix} \begin{bmatrix} 1 & 1 \\ 1 & 1 \\ 1 & 1 \\ 0 & 0 \\ 1 & 1 \\ 1 & 1 \\ 1 & 0 \\ 1 & 1 \end{bmatrix} = \begin{bmatrix} 0 & 0 \end{bmatrix}.$$

This result indicates that $\langle r_{13}| = \begin{bmatrix} 0 & 0 \end{bmatrix}$, therefore the polarity-6 RM expansion for the switching algebra model of the network does not include the product term $\bar{x}_1 x_3$ in either $f(x_1, x_2, x_3)$ or $\bar{f}(x_1, x_2, x_3)$. □

The computation of single RM spectral coefficients using the spectral response matrix is easily extended to the case of computing subsets of coefficients. To compute a subset of q RM spectral coefficients for an n-input, m-output switching network characterized by an RM spectral response matrix, $\mathbf{S}_f$, an index matrix $\mathbf{Q}_6$ of size $q \times 2^m$ is formed where each row vector in $\mathbf{Q}_6$ points to one of the desired coefficients within $\mathbf{S}_f$.

Example 6.26 *Subsets of RM Spectral Coefficient*
If it is desired to compute the polarity-6 RM spectral coefficients $\langle r_0|$, $\langle r_{12}|$, and $\langle r_13|$ for the example $n = 3$ primary input, $m = 1$ primary output switching network used in Examples 6.24 and 6.25, the index matrix $\mathbf{Q}_0$ is first formulated for the polarity-0 RM spectral response case.

$$\mathbf{Q}_0 = \begin{bmatrix} \langle 000| \\ \langle 110| \\ \langle 101| \end{bmatrix}$$

Next, the row vectors of $\mathbf{Q}_0$ are permuted to account for the naturally ordered polarity-6 spectral response matrix by performing a bitwise exclusive-OR operation with each index row vector comprising $\mathbf{Q}_0$ with the binary value of the polarity value, $6_{10} = 110_2$ resulting in $\mathbf{Q}_6$.

$$\mathbf{Q}_6 = \begin{bmatrix} \langle(000) \oplus (110)| \\ \langle(110) \oplus (110)| \\ \langle(101) \oplus (110)| \end{bmatrix} = \begin{bmatrix} \langle 110| \\ \langle 000| \\ \langle 011| \end{bmatrix} = \begin{bmatrix} 0 & 0 & 0 & 0 & 0 & 0 & 1 & 0 \\ 1 & 0 & 0 & 0 & 0 & 0 & 0 & 0 \\ 0 & 0 & 0 & 1 & 0 & 0 & 0 & 0 \end{bmatrix}$$

The desired subset of polarity-6 RM spectral coefficients may now be calculated using the index matrix $\mathbf{Q}_6$ and the polarity-6 spectral response matrix.

$$
\begin{bmatrix} \langle r_0| \\ \langle r_{12}| \\ \langle r_{13}| \end{bmatrix} = \mathbf{Q}_f \mathbf{S}_f = \begin{bmatrix} 0 & 0 & 0 & 0 & 0 & 0 & 1 & 0 \\ 1 & 0 & 0 & 0 & 0 & 0 & 0 & 0 \\ 0 & 0 & 0 & 1 & 0 & 0 & 0 & 0 \end{bmatrix} \begin{bmatrix} 1 & 1 \\ 1 & 1 \\ 1 & 1 \\ 0 & 0 \\ 1 & 1 \\ 1 & 1 \\ 1 & 0 \\ 1 & 1 \end{bmatrix} = \begin{bmatrix} 1 & 0 \\ 1 & 1 \\ 0 & 0 \end{bmatrix}
$$

In terms of the switching algebra polarity-6 RM expansion, the result of this calculation indicates that the product terms 1 and $\bar{x}_1 \bar{x}_2$ are included in the expansion for $\bar{f}(x_1, x_2, x_3)$ and the product term $\bar{x}_1 \bar{x}_2$ is included in the expansion for $f(x_1, x_2, x_3)$. Since $\langle r_{13}| = \begin{bmatrix} 0 & 0 \end{bmatrix}$, the corresponding product term $\bar{x}_1 x_3$ is not included in the expansions for either $\bar{f}(x_1, x_2, x_3)$ or $f(x_1, x_2, x_3)$. $\qquad\square$

6.5.4   CALCULATION OF THE RM SPECTRA THROUGH NETLIST TRAVERSALS

As previously described, the transfer matrix characterizing a switching network may be expressed in distributed factored form and represented by a set of interconnected transfer matrices that represent the network elements. This form of the transfer function when represented graphically corresponds to the topology of the switching network netlist. Since the RM spectrum of a switching network can be computed using the transfer matrix, the use of the distributed factored form for the computation of the RM spectrum allows a process to be formulated that computes coefficients through a netlist traversal.

To demonstrate this process, the switching network depicted in Figure 6.5 is used as an example and is presented again in Figure 6.12a with a graphical representation of the distributed factored form below it in Figure 6.12b. The fanout node in the distributed factored form diagram is replaced with $\mathbf{I}$ instead of $\mathbf{FO}$ to simplify the RM coefficient computation through netlist traversal. As described in the previous section for Walsh spectral coefficient calculations, the use of $\mathbf{FO}$ is required when the monolithic form of the transfer matrix is computed; however, for the purposes of traversing a distributed factored form, using $\mathbf{I}$ to model fanout points simplifies the process.

The RM spectral response matrix can also be graphically depicted in distributed factored form and that diagram is present in Figure 6.12c. Definition 6.4 specifies the RM spectral response matrix as

$$\mathbf{S}_f = \mathbf{R}_n \mathbf{T}_f.$$

The example network in Figure 6.5 is a $n = 2$ primary input, $m = 2$ primary output network, thus $\mathbf{R}_2$ is the appropriate RM spectral transformation matrix to be used in the determination of the corresponding spectral response matrix. Since $\mathbf{R}_2 = \mathbf{R}_1 \otimes \mathbf{R}_1$, the polarity-0 distributed factored form for the spectral response matrix is graphically depicted by including the $\mathbf{R}_1$ matrix at each network input rather than $\mathbf{R}_2$. This aids and facilitates the netlist traversal method for computing RM spectral coefficients. If it is desired to represent the RM spectral response matrix for a non-zero polarity, $\mathbf{R}_1$ and $\mathbf{R}_{n1}$ are used at the appropriate network inputs in accordance with Example 6.13.

Figure 6.12d depicts the distributed factored form of polarity-0 RM spectral response matrix with the inputs, outputs, and internal lines annotated with vector values corresponding to the computation of the r_0 coefficient. The calculation is performed by annotating the primary inputs with the appropriate index vector values. In this case, the r_0 coefficient is computed from the spectral response matrix by multiplying with the $\langle 00|$ index vector. Since the traversal of the netlist requires individual assignment of vectors in $\mathbb{H}$ at each primary input, the index vector is factored as $\langle 00| = \langle 0| \otimes \langle 0|$ and $\langle 0|$ is assigned to each primary input. These values are then propagated toward the primary outputs in a depth first fashion by multiplying them with encountered transfer matrices for each network operator. When an operator is encountered that has more than a single input, the vector values present at each input are combined into a single vector using the outer product operator. The result of the traversal depicted in Figure 6.12d indicates that $\langle r_0| = \langle 1|$ for output $\langle f_1|$ and $\langle r_0| = \langle 0|$ for output $\langle f_2|$.

The other three polarity-0 RM spectral coefficients are computed in the distributed factored form of the spectral response matrix in Figure 6.13a–c. The results for primary output $\langle f_1|$ are $\langle r_2| = \langle 1|$, $\langle r_1| = \langle 1|$, and $\langle r_{12}| = \langle 1|$. The results for primary output $\langle f_2|$ are $\langle r_2| = \langle 1|$, $\langle r_1| = \langle 1|$, and $\langle r_{12}| = \langle 1|$.

To verify the results of the netlist traversals, the polarity-0 spectral coefficients are calculated for each output in the form of the spectral response matrices for each output. The form of each polarity-0 spectral response matrix is

$$
\mathbf{S}_f = \begin{bmatrix} \langle r_0| \\ \langle r_2| \\ \langle r_1| \\ \langle r_{12}| \end{bmatrix}.
$$

The spectral response matrices $\mathbf{S}_{f_1}$ and $\mathbf{S}_{f_2}$ are

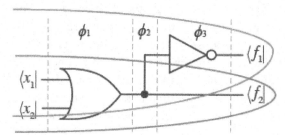

a) Example Network with Partitions and Cones

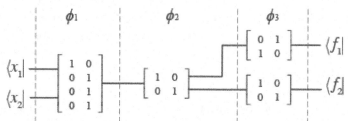

b) Transfer Matrix in Distributed Factored Form

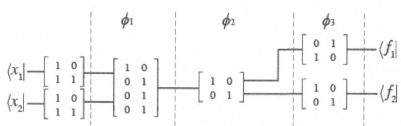

c) Polarity-0 Spectral Response Matrix in Distributed Factored Form

d) Polarity-0 Spectral Response Matrix in Distributed Factored Form
Annotated with RM Spectral Coefficient Values for r_0

Figure 6.12: Distributed factored form a) transfer matrix b) RM polarity-0 spectral response matrix.

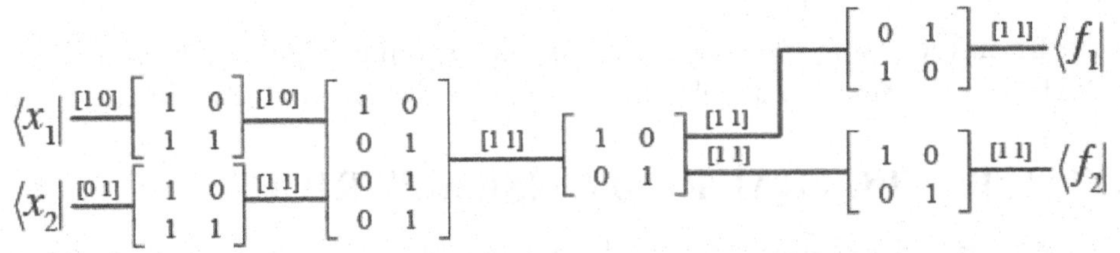

a) Polarity-0 Spectral Response Matrix in Distributed Factored Form Annotated with RM Spectral Coefficient Values for r_2

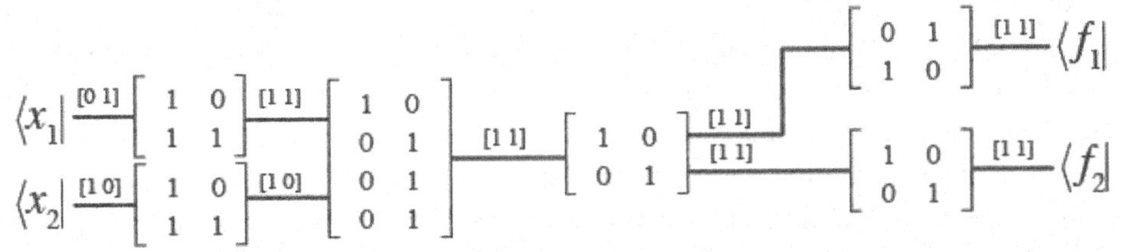

b) Polarity-0 Spectral Response Matrix in Distributed Factored Form Annotated with RM Spectral Coefficient Values for r_1

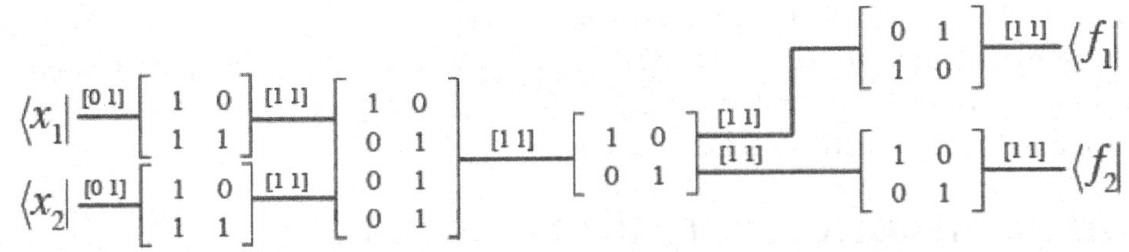

c) Polarity-0 Spectral Response Matrix in Distributed Factored Form Annotated with RM Spectral Coefficient Values for r_{12}

Figure 6.13: Distributed factored form RM spectral computations for a) r_2 b) r_1 c) r_{12}.

$$S_{f_1} = R_2 T_{f_1} = \begin{bmatrix} 1 & 0 & 0 & 0 \\ 1 & 1 & 0 & 0 \\ 1 & 0 & 1 & 0 \\ 1 & 1 & 1 & 1 \end{bmatrix} \begin{bmatrix} 0 & 1 \\ 1 & 0 \\ 1 & 0 \\ 1 & 0 \end{bmatrix} = \begin{bmatrix} 0 & 1 \\ 1 & 1 \\ 1 & 1 \\ 1 & 1 \end{bmatrix}$$

$$S_{f_2} = R_2 T_{f_2} = \begin{bmatrix} 1 & 0 & 0 & 0 \\ 1 & 1 & 0 & 0 \\ 1 & 0 & 1 & 0 \\ 1 & 1 & 1 & 1 \end{bmatrix} \begin{bmatrix} 1 & 0 \\ 0 & 1 \\ 0 & 1 \\ 0 & 1 \end{bmatrix} = \begin{bmatrix} 1 & 0 \\ 1 & 1 \\ 1 & 1 \\ 1 & 1 \end{bmatrix}$$

As can be observed from the row vectors in each spectral response matrix, the correct polarity-0 RM spectral coefficients are obtained from the netlist traversal procedure.

6.6   OTHER SPECTRAL RESPONSE MATRICES

The previous sections in this chapter introduced the notion of and derived spectral response matrices that characterize a switching network in the Walsh and RM spectral domains. It is shown that the spectral response matrices can be used to compute network output response in either the spectral or switching domains through a direct vector-matrix multiplication and that, furthermore, the spectral response matrix can be used to determine the complete spectra, a subset of spectral coefficients, or a single spectral coefficient. Because the transfer matrix of a switching network can be expressed in distributed factored form, a spectral coefficient can be computed through a traversal of the structural netlist description of the network. This is a significant result because it allows spectral coefficients to be computed based on netlist descriptions without the requirement of first extracting and representing a monolithic model of the underlying switching functions—a problem that has plagued and prevented widespread usage of spectral methods in modern EDA software tools.

The methods described in the previous chapter sections are based upon the Walsh and RM spectra. However, it is noted that these ideas and methods may be analogously developed and applied to any arbitrary spectral transformation with the only requirement being that the underlying spectral transformation matrix is orthogonal.

6.6.1   THE ARITHMETIC SPECTRAL TRANSFORM

An example of another applicable discrete orthogonal transformation is the *arithmetic spectrum* [50]. The arithmetic transformation matrix $\mathbf{A}_n$ is given in Equation 6.7 and is described in more detail in [7]. This transform is also known as the *probability transform* [51] and the *inverse integer Reed-Muller transform* [22]. Originally the arithmetic transform was introduced in [53], although it was not referred to by this name.

$$\mathbf{A}_1 = \begin{bmatrix} 1 & 0 \\ -1 & 1 \end{bmatrix} \qquad\qquad \mathbf{A}_n = \bigotimes_{i=1}^{n} \mathbf{A}_1 \qquad (6.7)$$

Like the Walsh transform, and unlike the RM transform, addition operations are carried out over the field of integers (not $GF(2)$). Therefore the inverse transform is given as

$$\mathbf{A}_n^{-1} = \bigotimes_{i=1}^{n} \begin{bmatrix} 1 & 0 \\ 1 & 1 \end{bmatrix},$$

thus leading to the alternative name *inverse integer Reed-Muller transform*.

CHAPTER 7

Multi-valued Switching Network Spectra

7.1 MV SPECTRA

The results of the previous chapter are easily generalized to the case of finite multi-valued switching networks and their corresponding linear algebraic models. Without loss of generality, we focus upon the example of a ternary MVSN although the results are applied for any radix $r > 2$. For brevity, the generalization of the Walsh transform is used in this chapter; however any discrete orthogonal spectral transformation matrix can be used with similar results.

7.2 CHRESTENSON TRANSFORM

The Chrestenson transform is a generalization of the Walsh transform. The Walsh basis functions are periodic functions that have two values equivalent to the square roots of unity, $\{-1, +1\}$. The set of basis functions comprising the Chrestenson transform are likewise periodic functions with respect to a parameter r where $r > 2$ [54]. The Chrestenson basis functions have r distinct values that are the rth roots of unity. When $r > 2$, the rth roots of unity are complex-valued and lie along the unit circle in the complex plane. The Chrestenson transform is then a Fourier transform parameterized for discrete finite-valued functions by their radix value, r and is applicable for applications to MVSNs.

As described in Chapter 5, MVSNs are traditionally modeled with scalar-valued switching algebras that are often generalizations or supersets of Boolean algebras to provide functional completeness such as the algebras proposed by Emil Post [30]. Many specific examples of different scalar-valued algebraic systems have been proposed and used in the past and some of these were motivated by choosing a collection of operators that correspond to the functionality of particular electronic components or subcircuits. However, these are scalar-valued switching algebras and thus the spectral analysis of these systems result in scalar-valued Chrestenson spectral coefficients. In this chapter, we introduce the vector-valued Chrestenson transform that comports with the vector space model of MVSNs described in Chapter 5.

For illustrative purposes, we shall focus our discussion upon the Chrestenson spectrum of a ternary-valued $r = 3$ MVSN that can be modeled using conventional switching functions that have integral values $\{0, 1, 2\}$ although these concepts are applicable to higher radix systems with alternative functional valuation sets. In general, the results provided in this chapter are applicable to any discrete-valued MVSN with $r > 2$ distinct valuations. Conventionally, this class of MVSNs are modeled with a switching algebra comprised of a set of elements from the non-negative integers, $\{0, 1, \ldots, r - 1\}$, although other valuation sets are possible and can be used. In this chapter, we assume that multiple-valued switching functions are those modeled with Postian algebra and have domain and range sets comprised of elements that are non-negative integers unless specified otherwise. We also assume that ternary MVSNs can be modeled as interconnected graphs of operators or gates that are depicted in Figure 5.3 in Chapter 5. Such graphs are colloquially referred to as "logic" or "circuit" diagrams since MVSNs are often physically realized as electronic circuits. A textual description of such an interconnected graph of gates or operators is referred to as a "netlist." Netlists are commonly used within the circuit design community and typically adhere to the specification and syntax of some hardware description language that is capable of such a structural representation. Therefore, a structural netlist and a circuit diagram representation of a given MVSN are equivalent. We use an extension of the IEEE standardized HDL to describe structural netlist descriptions of MVSNs comprised of the operators depicted in Figure 5.3. Such an extension of the Verilog HDL has been introduced in the work [55] based on SystemVerilog.

7.2.1   CHRESTENSON TRANSFORM OF SCALAR-VALUED SWITCHING FUNCTIONS

A ternary-valued discrete function has a Chrestenson spectral representation corresponding to a weighted set of discrete basis functions with valuations of the cube roots of unity, $\{a_0, a_1, a_2\}$ where $a_0^3 = a_1^3 = a_2^3 = 1$ and a_0 is defined as $a_0 = 1$. These cube roots of unity are graphically depicted as values lying upon the unit circle in the complex plane in Figure 6.1.

Initially, a mapping of the switching function values of $\{0, 1, 2\} \rightarrow \{a_0, a_1, a_2\}$ occurs before transformation to the Chrestenson spectral domain as in [39] [56]. For the purposes of differentiating which of these two functional valuation sets is being used, we refer to the use of $\{0, 1, 2\}$ as *scalar switching values* and the use of $\{a_0, a_1, a_2\}$ as *scalar complex values*. The mapping of the scalar r-valued switching constants to their corresponding complex valued counterparts is accomplished via Equation 7.1 where $k \in \{0, 1, \ldots, r - 1\}$.

$$a_k = e^{i\frac{2k\pi}{r}} \tag{7.1}$$

Although it is possible to interpret a complex value as a two-dimensional vector in the complex plane, we use the term *scalar complex value* because these complex values weight or scale the Chrestenson basis functions after spectral transformation. Additionally, we define another set of values known as the *vector switching values* and *vector complex values* in a later section of this chapter. Table 7.1 contains the correspondence of the scalar switching and scalar complex ternary constant values.

The three cube roots of unity are geometrically spaced at 120° angles along the boundary of a unit circle in the complex plane with a_0 located on the real axis at 0°. There are several useful

Table 7.1: Scalar ternary logic constants

Scalar Ternary Values	
Switching	Complex
0	a_0
1	a_1
2	a_2

relationships among these values that can aid in performing calculations. These properties are listed below.

Cube Root of Unity Properties		
$a_0 = 1$	$a_1 = -\frac{1}{2} + i\frac{\sqrt{3}}{2}$	$a_2 = -\frac{1}{2} - i\frac{\sqrt{3}}{2}$
$a_0 \times a_0 = a_0$	$a_0 \times a_1 = a_1$	$a_0 \times a_2 = a_2$
$a_1 \times a_1 = a_2$	$a_2 \times a_2 = a_1$	$a_2 \times a_1 = a_0$
$a_0 + a_1 + a_2 = 0$	$a_1 + a_2 = -1$	$a_1 - a_2 = i\sqrt{3}$
$a_0^* = a_0$	$a_1^* = a_2$	$a_2^* = a_1$

Definition 7.1 *Scalar Complex Encoded Switching Function*

A ternary switching function that models a ternary MVSN, f, is in scalar complex encoded form, denoted as f_c, when all discrete scalar switching values are replaced with scalar complex values in accordance with Table 7.1. □

The Chrestenson transform is a discrete orthogonal transform and can thus be characterized by a spectral transformation matrix that is multiplied with a vector whose components are all possible valuations of discrete scalar switching function of interest. The column vector representing the switching function of interest is comprised of components that are the complex scalar encoded function range values. The transformation matrix can be constructed in natural order through the use of the outer product operation. Equations 7.2 define the naturally ordered Chrestenson spectral transformation matrix $\mathbf{C}_n$ for any $n \in \mathbb{N}$.

$$\mathbf{C}_1 = \begin{bmatrix} a_0 & a_0 & a_0 \\ a_0 & a_1 & a_2 \\ a_0 & a_2 & a_1 \end{bmatrix} \qquad \mathbf{C}_n = \bigotimes_{i=1}^{n} \mathbf{C}_1 \qquad (7.2)$$

The scalar complex encoded Chrestenson spectrum of a logic network is a column vector of spectral coefficients $|s_f\rangle$ and is computed through multiplication of $\mathbf{C}_m^*$ with a column vector of the complex scalar encoded values of the function f denoted as $|f_c\rangle$. Equation 7.3 expresses this relationship as used in [39] [56].

$$|s_f\rangle = \mathbf{C}_m^* |f_c\rangle \tag{7.3}$$

Example 7.2 *Chrestenson Spectrum of Scalar Encoded Switching Function*
Consider the J_2 (literal selection) gate shown in Figure 5.3. Table 7.2 contains the switching function model in tabular form for this network element with both scalar switching and scalar complex-values.

Table 7.2: Scalar ternary logic constants

Input Value	Switching Scalar Output Value	Complex Scalar Output Value
0	0	a_0
1	0	a_0
2	2	a_2

Since the number of primary inputs is one, $n = 1$, the appropriate spectral transformation matrix is $\mathbf{C}_1$. The spectrum is calculated as the direct product of the Hermitian (conjugate transpose) of $\mathbf{C}_1$ denoted as $\mathbf{C}_1^*$ with a column vector $|J_{2c}\rangle$ composed of all possible complex scalar output values for the J_2 gate. The Hermitian is

$$\mathbf{C}_1^* = \begin{bmatrix} a_0 & a_0 & a_0 \\ a_0 & a_1 & a_2 \\ a_0 & a_2 & a_1 \end{bmatrix}^* = \begin{bmatrix} a_0 & a_0 & a_0 \\ a_0 & a_2 & a_1 \\ a_0 & a_1 & a_2 \end{bmatrix}$$

The scalar Chrestenson spectrum for the J_2 gate is computed as:

$$|c_{J_2}\rangle = \mathbf{C}_1^* |J_{2c}\rangle = \begin{bmatrix} a_0 & a_0 & a_0 \\ a_0 & a_2 & a_1 \\ a_0 & a_1 & a_2 \end{bmatrix} \begin{bmatrix} a_0 \\ a_0 \\ a_2 \end{bmatrix} = \begin{bmatrix} 2a_0 + a_2 \\ 2a_0 + a_2 \\ a_0 + 2a_1 \end{bmatrix}$$

$\square$

7.2.2 CHRESTENSON TRANSFORM OF VECTOR-VALUED SWITCHING FUNCTIONS

Ternary MVSNs are modeled in the vector space by using three-dimensional canonical basis vectors to represent each switching constant value. To comport with the method of computing the Chrestenson spectrum of scalar-valued functions, we define vector-valued complex constants that serve as mapped counterparts to vector-valued switching constants in Table 7.3. Because we define these constants as row vectors, it is necessary to represent them as the conjugate transpose of their more intuitive column vector definition.

Table 7.3: Vector ternary switching constants

Vector Ternary Values		
Switching	Complex	
$\langle 0 \vert = \begin{bmatrix} 1 & 0 & 0 \end{bmatrix}$	$\langle c_0 \vert = \begin{bmatrix} a_0 & a_0 & a_0 \end{bmatrix}$	
$\langle 1 \vert = \begin{bmatrix} 0 & 1 & 0 \end{bmatrix}$	$\langle c_1 \vert = \begin{bmatrix} a_0 & a_2 & a_1 \end{bmatrix}$	
$\langle 2 \vert = \begin{bmatrix} 0 & 0 & 1 \end{bmatrix}$	$\langle c_2 \vert = \begin{bmatrix} a_0 & a_1 & a_2 \end{bmatrix}$	

The vector Chrestenson spectrum $\mathbf{S}_f$ is calculated using the relationship in Equation 7.3 where the column vector of scalar complex conjugate encodings for the function $\vert f_c \rangle$ is replaced with matrix $\mathbf{T}_s$. Each row of $\mathbf{T}_s$ is the vector complex encoding of f values, $\langle f_c \vert$, and $\mathbf{T}$ is the switching domain transfer matrix. By the property of truth table isomorphism, the transfer matrix $\mathbf{T}$ can be viewed as a single column of row vectors where each row vector is the switching vector encoded truth value of function f, hence Equation 7.3 yields $\mathbf{T}_s$, the complex vector encoded form of $\mathbf{T}$. This observation leads to Definition 7.3.

Definition 7.3   *Chrestenson Spectral Response Matrix*
The Chrestenson spectral response matrix $\mathbf{T}_s$ models the functionality of a MVSN in the Chrestenson spectral domain. The Chrestenson spectral response matrix is defined in Equation 7.3.

$$\mathbf{T}_s = \mathbf{T}\mathbf{C}_n^* \tag{7.4}$$

□

Using Equations 7.3 and 7.4, the vector Chrestenson spectrum $\mathbf{S}_f$ of the n-input, m-output MVSN modeled by transfer matrix T is given in Equation 7.5.

$$\mathbf{S}_f = \mathbf{C}_m^* \mathbf{T}_s \tag{7.5}$$

Substituting Equation 7.4 in Equation 7.5, the relationship between the switching domain transfer matrix and the Chrestenson spectrum of an n-input, m-output MVSN becomes

$$\mathbf{S}_f = \mathbf{C}_m^* \mathbf{T}\mathbf{C}_n^*.$$

Example 7.4 *Vector Chrestenson Spectrum* To compute the vector Chrestenson spectrum of the J_2 gate, we first formulate the vector complex encoded values of J_2 as the matrix $\mathbf{J}_{2s}$ through the application of the mapping in Equation 7.4.

$$\mathbf{J}_{2s} = (\mathbf{J}_2)(\mathbf{C}_1^*) = \begin{bmatrix} 1 & 0 & 0 \\ 1 & 0 & 0 \\ 0 & 0 & 1 \end{bmatrix} \begin{bmatrix} a_0 & a_0 & a_0 \\ a_0 & a_2 & a_1 \\ a_0 & a_1 & a_2 \end{bmatrix} = \begin{bmatrix} a_0 & a_0 & a_0 \\ a_0 & a_0 & a_0 \\ a_0 & a_1 & a_2 \end{bmatrix}$$

The calculation of the spectral matrix $\mathbf{S}_{J2}$ is then accomplished using Equation 7.5.

$$\mathbf{S}_{J2} = (\mathbf{C}_1^*)(\mathbf{J}_{2s}) = \begin{bmatrix} a_0 & a_0 & a_0 \\ a_0 & a_2 & a_1 \\ a_0 & a_1 & a_2 \end{bmatrix} \begin{bmatrix} a_0 & a_0 & a_0 \\ a_0 & a_0 & a_0 \\ a_0 & a_1 & a_2 \end{bmatrix}$$

$$= \begin{bmatrix} (3a_0) & (2a_0 + a_1) & (2a_0 + a_2) \\ (a_0 + a_1 + a_2) & (a_0 + 2a_2) & (2a_0 + a_2) \\ (a_0 + a_1 + a_2) & (2a_0 + a_1) & (a_0 + 2a_1) \end{bmatrix}$$

□

Each individual Chrestenson spectral coefficient is a row vector within the $\mathbf{S}_f$ spectral matrix, and several properties of these coefficients are apparent. The $0^t h$-ordered coefficient is the topmost row vector of $\mathbf{S}_f$ and the first component of this coefficient is always the real value r^n where r is the radix or number of distinct function switching values and n is the number of primary inputs of the MVSN. Furthermore, the first component of the remaining higher-ordered spectral coefficients is always zero.

The rightmost component of each spectral coefficient vector is identical to the scalar Chrestenson spectrum when the MVSN is modeled in the switching algebra domain rather than the vector space domain.

Theorem 7.5 Scalar and Vector Spectrum Relation
The scalar Chrestenson spectrum $|s_f\rangle$ and the rightmost column of the vector Chrestenson spectrum $\mathbf{S}_f$ of a function representing the same ternary MVSN are identical.

Proof. For a given switching function f, the scalar Chrestenson spectrum is given as $|s_f\rangle = \mathbf{C}_n^* | f_c\rangle$ and the vector Chrestenson spectrum is given as $\mathbf{S}_f = \mathbf{C}_n^* \mathbf{T}_s$. We define a vector $|v\rangle$ of length 3^n composed of $3^n - 1$ zero-valued components and a single unity-valued component of the form, $\langle v| = \begin{bmatrix} 0 & 0 & \dots & 0 & 1 \end{bmatrix}$. The rightmost column of the vector Chrestenson spectrum can be formed by the product $\mathbf{S}_f |v\rangle$, yielding:

Figure 7.1: Diagram of transfer and spectral response matrices.

$$\mathbf{S}_f|v\rangle = \mathbf{C}_n^* \mathbf{T}_s|v\rangle \tag{7.6}$$

Subtracting Equation 7.3 from Equation 7.6 results in:

$$\mathbf{S}_f|v\rangle - |s_f\rangle = \mathbf{C}_n^* \mathbf{T}_s|v\rangle - \mathbf{C}_n^*|f_c\rangle$$
$$= \mathbf{C}_n^*(\mathbf{T}_s|v\rangle - |f_c\rangle) \tag{7.7}$$

For a given switching function f, $\mathbf{T}_s$ and $|f_c\rangle$ result by expressing the components of $|f\rangle$ using the vector complex and scalar complex encoded values from Tables 7.1 and 7.3. Examination of the encodings specified in Tables 7.1 and 7.3 reveals that the scalar complex and vector complex encodings are identical in that they both have value a_i corresponding to f having a scalar switching value of i. Thus, the Equation 7.7 term $(\mathbf{T}_s|v\rangle - |f_c\rangle) = |\varnothing\rangle$. Substituting this observation into Equation 7.7 results in $\mathbf{S}_f|v\rangle - |s_f\rangle = \mathbf{C}_n^*|\varnothing\rangle = |\varnothing\rangle$. This result can only occur if the rightmost column of $\mathbf{S}_f$ is equivalent to $|s_f\rangle$. □

7.2.3   CHRESTENSON SPECTRAL RESPONSE MATRICES

The spectral response matrix $\mathbf{T}_s$ is a form of MVSN transfer matrix that can provide both the Chrestenson spectral response or the output switching response depending upon the form of the input stimulus vector used to multiply or excite $\mathbf{T}_s$. Figure 7.1 contains a block diagram where a ternary logic network is represented by a spectral transfer matrix $\mathbf{T}_s$ with an input stimulus matrix $\mathbf{X}_c$ and a spectral response matrix $\mathbf{S}_f$.

Theorem 7.6   Spectral Transfer Matrix
The spectral transfer matrix $\mathbf{T}_s$ for a logic network is related to the switching transfer matrix $\mathbf{T}$ as given in Equation 7.8.

$$\mathbf{T}_s = \mathbf{T}\mathbf{C}_m^* \tag{7.8}$$

Proof. Substituting Equation 7.4 into Equation 7.5 results in:

$$\mathbf{S}_f = \mathbf{C}_n^* \mathbf{F} \mathbf{C}_m^* \tag{7.9}$$

Using the notion of a switching transfer matrix, the switching domain output response, $\langle f|$, due to a single switching vector input stimulus $\langle x|$ is computed as $\langle f| = \langle x|\mathbf{T}$. If all possible valuations $\langle x_i|$ (where $0 < i < 3^n - 1$) are represented as a column of row vectors, a matrix $\mathbf{X}$ results with each row equivalent to $\langle x_i|$. Using $\mathbf{X}$ to compute the total switching vector response results in $\mathbf{F} = \mathbf{X}\mathbf{T}$. Furthermore, due to the switching vector encoding definition given in Tables 7.1 and 7.3, it is observed that $\mathbf{X} = \mathbf{I}$ where $\mathbf{I}$ is the identity matrix, hence $\mathbf{F} = \mathbf{X}\mathbf{T} = \mathbf{I}\mathbf{T} = \mathbf{T}$.

Substituting this result into Equation 7.9 results in:

$$\mathbf{S}_f = \mathbf{C}_n^* \mathbf{X} \mathbf{T} \mathbf{C}_m^* = \mathbf{C}_n^* \mathbf{T} \mathbf{C}_m^* \tag{7.10}$$

It is observed that the leftmost $\mathbf{C}_n^*$ factor in the expression $\mathbf{C}_n^* \mathbf{T} \mathbf{C}_m^*$ can be considered to be composed of a single column of row vectors where each row vector represents the vector complex encoded input values $\mathbf{X}_c$. From this observation, we have $\mathbf{S}_f = \mathbf{X}_c \mathbf{T} \mathbf{C}_m^*$. Hence, $\mathbf{T}_s = \mathbf{T} \mathbf{C}_m^*$.
□

Subsets or Single Output Response using the Spectral Response Matrix

The Chrestenson spectral response matrix can be used to calculate single or subsets of output responses resulting from a specified set of input stimuli. The Chrestenson spectrum of an MVSN is a complete characterization of the network in the spectral domain. Another viewpoint is to consider a single spectral coefficient as a network output response due to a specific input stimulus in the Chrestenson spectrum domain. As is the case with the spectral response matrices for binary-valued networks described previously, the Chrestenson spectral response matrix can be used to compute output responses in either the spectral or the switching domains.

To compute the spectral response for a single input stimulus vector, Equation 7.5 is modified and is of the form

$$\langle s_x| = \langle x_s|\mathbf{T}_s$$

where $\langle x_s|$ represents an input stimulus vector expressed in the Chrestenson domain and $\langle s_x|$ is the resulting spectral output response of the MVSN. Likewise, the switching output response can be obtained by

$$\langle f_c| = \langle x|\mathbf{T}_s$$

where the input stimulus $\langle x|$ is specified as a vector-valued switching model for an MVSN input excitation, and the corresponding output response is the vector complex encoded MVSN output response.

Example 7.7   *MVSN Single Output Response using the Spectral Response Matrix*
Consider an MVSN consisting of a single J_2 literal gate. The spectral output response for an input stimulus of $\langle x| = \langle 2|$ is computed by first specifying the input stimulus in the Chrestenson spectral domain.

$$\langle x_s| = \langle x_c|\mathbf{C}_1^* = \begin{bmatrix} 0 & 0 & 1 \end{bmatrix} \begin{bmatrix} a_0 & a_0 & a_0 \\ a_0 & a_2 & a_1 \\ a_0 & a_1 & a_2 \end{bmatrix} = \begin{bmatrix} a_0 & a_1 & a_2 \end{bmatrix}$$

Next, $\langle x_s|$ is multiplied with the spectral response matrix $\mathbf{J}_{2s}$ yielding the spectral output response.

$$\langle f_s| = \langle x_s|\mathbf{J}_{2s} = \begin{bmatrix} a_0 & a_1 & a_2 \end{bmatrix} \begin{bmatrix} a_0 & a_0 & a_0 \\ a_0 & a_0 & a_0 \\ a_0 & a_1 & a_2 \end{bmatrix} \begin{bmatrix} a_0 & a_1 & a_2 \end{bmatrix}$$
$$= \begin{bmatrix} (a_0 + a_1 + a_2) & (2a_0 + a_1) & (a_0 + 2a_1) \end{bmatrix}$$

The Chrestenson spectral response matrix may also be used to compute the output response in the switching domain by multiplying the input stimulus $\langle x| = \langle 2|$ directly with the MVSN spectral response matrix.

$$\langle f_c| = \langle x|\mathbf{J}_{2s} = \begin{bmatrix} 0 & 0 & 1 \end{bmatrix} \begin{bmatrix} a_0 & a_0 & a_0 \\ a_0 & a_0 & a_0 \\ a_0 & a_1 & a_2 \end{bmatrix}$$
$$= \begin{bmatrix} a_0 & a_1 & a_2 \end{bmatrix}$$

From Table 7.3, it is observed that $\langle f_c|$ is the vector complex value for $\langle 2|$ which is the expected output response of the J_2 literal selection gate when excited with input stimulus $\langle 2|$. If the output response is desired as a vector switching value, it can be further multiplied by $\frac{1}{3}\mathbf{C}_1$.

$$\begin{bmatrix} a_0 & a_1 & a_2 \end{bmatrix} \left(\frac{1}{3} \begin{bmatrix} a_0 & a_0 & a_0 \\ a_0 & a_1 & a_2 \\ a_0 & a_2 & a_1 \end{bmatrix} \right)$$
$$= \frac{1}{3} \begin{bmatrix} (a_0 + a_1 + a_2) & (a_0 + a_1 + a_2) & (a_0 + a_0 + a_0) \end{bmatrix}$$
$$= \begin{bmatrix} 0 & 0 & 1 \end{bmatrix} = \langle 2|$$

□

7.2.4   COMPUTING CHRESTENSON SPECTRAL COEFFICIENTS FROM A NETLIST

The Chrestenson spectral response matrix may be represented in distributed factored form as a graphical set of interconnected transfer matrices for each network element that has an identical topology to the MVSN with the addition of C_1^* matrices being present at each primary input. In this form, the output response in either the switching or the spectral domain may be computed through a traversal of the structure representing the distributed factored form of the spectral response matrix. Because the topologies of the MVSN and this form of the spectral response matrix are identical, this process allows spectral coefficients and MVSN simulation algorithms to be formulated through netlist traversals.

This technique is advantageous to previous approaches for spectral computations since past methods require the extraction of a complete algebraic switching function and the representation of both the Chrestenson spectral transformation matrix and the extracted switching function before the calculation of the spectrum could commence. Given that both the transformation matrix and the switching function are exponentially large with respect to n, the number of MVSN primary inputs, spectral methods have seen limited use in the past. The techniques enabled by the vector space model overcome these previous limitations since spectral coefficients can be computed through breadth-first traversals of graphs representing the distributed factored form of the spectral response matrix.

Example 7.8 computes the naturally ordered vector Chrestenson spectrum for the example MVSN depicted in Figure 5.3 in Chapter 5. The result of this calculation will be used to compare to the spectral coefficients computed through a traversal of the netlist.

Example 7.8   *Vector Chrestenson Spectrum Calculation using the Transfer Matrix*
The transfer matrix **T** is given as

$$
\mathbf{T} = \begin{bmatrix}
1 & 0 & 0 & 0 & 0 & 0 & 0 & 0 & 0 \\
0 & 0 & 0 & 0 & 0 & 0 & 0 & 1 & 0 \\
0 & 0 & 1 & 0 & 0 & 0 & 0 & 0 & 0 \\
0 & 0 & 0 & 0 & 0 & 0 & 0 & 1 & 0 \\
0 & 0 & 0 & 0 & 0 & 0 & 0 & 1 & 0 \\
0 & 0 & 1 & 0 & 0 & 0 & 0 & 0 & 0 \\
0 & 0 & 1 & 0 & 0 & 0 & 0 & 0 & 0 \\
0 & 0 & 1 & 0 & 0 & 0 & 0 & 0 & 0 \\
0 & 0 & 1 & 0 & 0 & 0 & 0 & 0 & 0
\end{bmatrix}.
$$

From Theorem 7.6, the vector Chrestenson spectrum is computed using the transfer matrix with Equation 7.9, requiring the formation of $\mathbf{C}_n^*$ and $\mathbf{C}_m^*$. In the case of the example MVSN, $n = m = 2$, therefore, the spectrum is computed as $\mathbf{S} = \mathbf{C}_2^* \mathbf{T} \mathbf{C}_2^*$. The matrix $\mathbf{C}_2^*$ in natural order is

$$\mathbf{C}_2^* = \begin{bmatrix} a_0 & a_0 & a_0 \\ a_0 & a_2 & a_1 \\ a_0 & a_1 & a_2 \end{bmatrix} \otimes \begin{bmatrix} a_0 & a_0 & a_0 \\ a_0 & a_2 & a_1 \\ a_0 & a_1 & a_2 \end{bmatrix}$$

$$= \begin{bmatrix} a_0 & a_0 & a_0 & a_0 & a_0 & a_0 & a_0 & a_0 & a_0 \\ a_0 & a_2 & a_1 & a_0 & a_2 & a_1 & a_0 & a_2 & a_1 \\ a_0 & a_1 & a_2 & a_0 & a_1 & a_2 & a_0 & a_1 & a_2 \\ a_0 & a_0 & a_0 & a_2 & a_2 & a_2 & a_1 & a_1 & a_1 \\ a_0 & a_2 & a_1 & a_2 & a_1 & a_0 & a_1 & a_0 & a_2 \\ a_0 & a_1 & a_2 & a_2 & a_0 & a_1 & a_1 & a_2 & a_0 \\ a_0 & a_0 & a_0 & a_1 & a_1 & a_1 & a_2 & a_2 & a_2 \\ a_0 & a_2 & a_1 & a_1 & a_0 & a_2 & a_2 & a_1 & a_0 \\ a_0 & a_1 & a_2 & a_1 & a_2 & a_0 & a_2 & a_0 & a_1 \end{bmatrix}.$$

The spectral response matrix is calculated as

$$\mathbf{T}_s = \mathbf{T}\mathbf{C}_2^* = \begin{bmatrix} 1 & 0 & 0 & 0 & 0 & 0 & 0 & 0 & 0 \\ 0 & 0 & 0 & 0 & 0 & 0 & 0 & 1 & 0 \\ 0 & 0 & 1 & 0 & 0 & 0 & 0 & 0 & 0 \\ 0 & 0 & 0 & 0 & 0 & 0 & 0 & 1 & 0 \\ 0 & 0 & 0 & 0 & 0 & 0 & 0 & 1 & 0 \\ 0 & 0 & 1 & 0 & 0 & 0 & 0 & 0 & 0 \\ 0 & 0 & 1 & 0 & 0 & 0 & 0 & 0 & 0 \\ 0 & 0 & 1 & 0 & 0 & 0 & 0 & 0 & 0 \\ 0 & 0 & 1 & 0 & 0 & 0 & 0 & 0 & 0 \end{bmatrix} \begin{bmatrix} a_0 & a_0 & a_0 & a_0 & a_0 & a_0 & a_0 & a_0 & a_0 \\ a_0 & a_2 & a_1 & a_0 & a_2 & a_1 & a_0 & a_2 & a_1 \\ a_0 & a_1 & a_2 & a_0 & a_1 & a_2 & a_0 & a_1 & a_2 \\ a_0 & a_0 & a_0 & a_2 & a_2 & a_2 & a_1 & a_1 & a_1 \\ a_0 & a_2 & a_1 & a_2 & a_1 & a_0 & a_1 & a_0 & a_2 \\ a_0 & a_1 & a_2 & a_2 & a_0 & a_1 & a_1 & a_2 & a_0 \\ a_0 & a_0 & a_0 & a_1 & a_1 & a_1 & a_2 & a_2 & a_2 \\ a_0 & a_2 & a_1 & a_1 & a_0 & a_2 & a_2 & a_1 & a_0 \\ a_0 & a_1 & a_2 & a_1 & a_2 & a_0 & a_2 & a_0 & a_1 \end{bmatrix}$$

$$= \begin{bmatrix} a_0 & a_0 & a_0 & a_0 & a_0 & a_0 & a_0 & a_0 & a_0 \\ a_0 & a_2 & a_1 & a_1 & a_0 & a_2 & a_2 & a_1 & a_0 \\ a_0 & a_1 & a_2 & a_0 & a_1 & a_2 & a_0 & a_1 & a_2 \\ a_0 & a_2 & a_1 & a_1 & a_0 & a_2 & a_2 & a_1 & a_0 \\ a_0 & a_2 & a_1 & a_1 & a_0 & a_2 & a_2 & a_1 & a_0 \\ a_0 & a_1 & a_2 & a_0 & a_1 & a_2 & a_0 & a_1 & a_2 \\ a_0 & a_1 & a_2 & a_0 & a_1 & a_2 & a_0 & a_1 & a_2 \\ a_0 & a_1 & a_2 & a_0 & a_1 & a_2 & a_0 & a_1 & a_2 \\ a_0 & a_1 & a_2 & a_0 & a_1 & a_2 & a_0 & a_1 & a_2 \end{bmatrix}.$$

$\mathbf{T}_s$ characterizes the transfer response of the example MVSN in the Chrestenson spectral domain. This matrix can be used to compute the complete Chrestenson spectrum $\mathbf{S}$ by premultiplying with $\mathbf{C}_2^*$ as follows.

$$\mathbf{S} = \mathbf{C}_2^* \mathbf{T}_s$$

$$= \begin{bmatrix}
a_0 & a_0 & a_0 & a_0 & a_0 & a_0 & a_0 & a_0 & a_0 \\
a_0 & a_2 & a_1 & a_0 & a_2 & a_1 & a_0 & a_2 & a_1 \\
a_0 & a_1 & a_2 & a_0 & a_1 & a_2 & a_0 & a_1 & a_2 \\
a_0 & a_0 & a_0 & a_2 & a_2 & a_2 & a_1 & a_1 & a_1 \\
a_0 & a_2 & a_1 & a_2 & a_1 & a_0 & a_1 & a_0 & a_2 \\
a_0 & a_1 & a_2 & a_2 & a_0 & a_1 & a_1 & a_2 & a_0 \\
a_0 & a_0 & a_0 & a_1 & a_1 & a_1 & a_2 & a_2 & a_2 \\
a_0 & a_2 & a_1 & a_1 & a_0 & a_2 & a_2 & a_1 & a_0 \\
a_0 & a_1 & a_2 & a_1 & a_2 & a_0 & a_2 & a_0 & a_1
\end{bmatrix}
\begin{bmatrix}
a_0 & a_0 & a_0 & a_0 & a_0 & a_0 & a_0 & a_0 & a_0 \\
a_0 & a_2 & a_1 & a_1 & a_0 & a_2 & a_2 & a_1 & a_0 \\
a_0 & a_1 & a_2 & a_0 & a_1 & a_2 & a_0 & a_1 & a_2 \\
a_0 & a_2 & a_1 & a_1 & a_0 & a_2 & a_2 & a_1 & a_0 \\
a_0 & a_2 & a_1 & a_1 & a_0 & a_2 & a_2 & a_1 & a_0 \\
a_0 & a_1 & a_2 & a_0 & a_1 & a_2 & a_0 & a_1 & a_2 \\
a_0 & a_1 & a_2 & a_0 & a_1 & a_2 & a_0 & a_1 & a_2 \\
a_0 & a_1 & a_2 & a_0 & a_1 & a_2 & a_0 & a_1 & a_2 \\
a_0 & a_1 & a_2 & a_0 & a_1 & a_2 & a_0 & a_1 & a_2
\end{bmatrix}.$$

Each spectral coefficient is of the form $s_0 a_0 + s_1 a_1 + s_2 a_2$. For conciseness, we use a triplet shorthand notation (s_0, s_1, s_2) to express each component in the spectral coefficient row vector. Using this notation, the total spectrum $\mathbf{S}$ is written as follows.

$$\mathbf{S} = \begin{bmatrix}
(9,0,0) & (1,5,3) & (1,3,5) & (6,3,0) & (4,5,0) & (1,0,8) & (6,0,3) & (1,8,0) & (4,0,5) \\
(3,3,3) & (2,3,4) & (6,2,1) & (4,4,1) & (3,1,5) & (4,3,2) & (2,5,2) & (4,2,3) & (5,1,3) \\
(3,3,3) & (6,1,2) & (2,4,3) & (2,2,5) & (5,3,1) & (4,3,2) & (4,1,4) & (4,2,3) & (3,5,1) \\
(3,3,3) & (2,3,4) & (6,2,1) & (4,4,1) & (3,1,5) & (4,3,2) & (2,5,2) & (4,2,3) & (5,1,3) \\
(3,3,3) & (3,4,2) & (5,1,3) & (5,2,2) & (2,3,4) & (4,3,2) & (4,4,1) & (4,2,3) & (3,2,4) \\
(3,3,3) & (4,2,3) & (4,3,2) & (3,3,3) & (4,2,3) & (4,3,2) & (3,3,3) & (4,2,3) & (4,3,2) \\
(3,3,3) & (6,1,2) & (2,4,3) & (2,2,5) & (5,3,1) & (4,3,2) & (4,1,4) & (4,2,3) & (3,5,1) \\
(3,3,3) & (4,2,3) & (4,3,2) & (3,3,3) & (4,2,3) & (4,3,2) & (3,3,3) & (4,2,3) & (4,3,2) \\
(3,3,3) & (5,3,1) & (3,2,4) & (4,1,4) & (3,4,2) & (4,3,2) & (5,2,2) & (4,2,3) & (2,4,3)
\end{bmatrix}$$

The spectral matrix may be simplified by applying the identities $a_0 + a_1 + a_2 = 0$ and $a_0 = 1$ to each component of $\mathbf{S}$.

$$\mathbf{S} = \begin{bmatrix}
9 & (0,4,2) & (0,2,4) & (6,3,0) & (4,5,0) & (1,0,8) & (6,0,3) & (1,8,0) & (4,0,5) \\
0 & (0,1,2) & (5,1,0) & (3,3,0) & (2,0,4) & (2,1,0) & (0,3,0) & (2,0,1) & (4,0,2) \\
0 & (5,0,1) & (0,2,1) & (0,0,3) & (4,2,0) & (2,1,0) & (3,0,3) & (2,0,1) & (2,4,0) \\
0 & (0,1,2) & (5,1,0) & (3,3,0) & (2,0,4) & (2,1,0) & (0,3,0) & (2,0,1) & (4,0,2) \\
0 & (1,2,0) & (4,0,2) & (3,0,0) & (0,1,2) & (2,1,0) & (3,3,0) & (2,0,1) & (1,0,2) \\
0 & (2,0,1) & (2,1,0) & 0 & (2,0,1) & (2,1,0) & 0 & (2,0,1) & (2,1,0) \\
0 & (5,0,1) & (0,2,1) & (0,0,3) & (4,2,0) & (2,1,0) & (3,0,3) & (2,0,1) & (2,4,0) \\
0 & (2,0,1) & (2,1,0) & 0 & (2,0,1) & (2,1,0) & 0 & (2,0,1) & (2,1,0) \\
0 & (4,2,0) & (1,0,2) & (3,0,3) & (1,2,0) & (2,1,0) & (3,0,0) & (2,0,1) & (0,2,1)
\end{bmatrix}$$

□

Each row vector comprising $\mathbf{S}$ is a vector Chrestenson spectral coefficient $\langle s_i |$ for the entire MVSN. If it is desired to compute the spectral coefficients of each individual primary output, cones are formed to partition the MVSN into those portions that only support the outputs

$\langle f_i |$. Figure 7.2 depicts the example network with partitioning cones. In Example 7.9, the vector Chrestenson spectrum is calculated for each of the individual MVSN primary outputs $\langle f_1 |$ and $\langle f_2 |$.

Example 7.9   *Vector Chrestenson Spectrum for MVSN Primary Outputs*

The technique described in Chapter 5 is used to form the switching transfer matrices for each portion of the example MVSN within a cone. These transfer matrices are computed as follows.

$$
\mathbf{T}_{f_1} = \mathbf{O}\mathbf{J}_1 =
\begin{bmatrix}
1 & 0 & 0 \\
0 & 1 & 0 \\
0 & 0 & 1 \\
0 & 1 & 0 \\
0 & 1 & 0 \\
0 & 0 & 1 \\
0 & 0 & 1 \\
0 & 0 & 1 \\
0 & 0 & 1
\end{bmatrix}
\begin{bmatrix}
1 & 0 & 0 \\
0 & 0 & 1 \\
1 & 0 & 0
\end{bmatrix}
=
\begin{bmatrix}
1 & 0 & 0 \\
0 & 0 & 1 \\
1 & 0 & 0 \\
0 & 0 & 1 \\
0 & 0 & 1 \\
1 & 0 & 0 \\
1 & 0 & 0 \\
1 & 0 & 0 \\
1 & 0 & 0
\end{bmatrix}
$$

$$
\mathbf{T}_{f_2} = \mathbf{O} =
\begin{bmatrix}
1 & 0 & 0 \\
0 & 1 & 0 \\
0 & 0 & 1 \\
0 & 1 & 0 \\
0 & 1 & 0 \\
0 & 0 & 1 \\
0 & 0 & 1 \\
0 & 0 & 1 \\
0 & 0 & 1
\end{bmatrix}
$$

Next, the spectral response matrices are formulated for each output as $\mathbf{T}_{sf_i} = \mathbf{T}_{f_i} \mathbf{C}_1^*$. The spectral response matrix for $\mathbf{T}_{sf_1}$ is

$$
\mathbf{T}_{sf_1} = \mathbf{T}_{f_1} \mathbf{C}_1^* =
\begin{bmatrix}
1 & 0 & 0 \\
0 & 0 & 1 \\
1 & 0 & 0 \\
0 & 0 & 1 \\
0 & 0 & 1 \\
1 & 0 & 0 \\
1 & 0 & 0 \\
1 & 0 & 0 \\
1 & 0 & 0
\end{bmatrix}
\begin{bmatrix}
a_0 & a_0 & a_0 \\
a_0 & a_2 & a_1 \\
a_0 & a_1 & a_2
\end{bmatrix}
=
\begin{bmatrix}
a_0 & a_0 & a_0 \\
a_0 & a_1 & a_2 \\
a_0 & a_0 & a_0 \\
a_0 & a_1 & a_2 \\
a_0 & a_1 & a_2 \\
a_0 & a_0 & a_0 \\
a_0 & a_0 & a_0 \\
a_0 & a_0 & a_0 \\
a_0 & a_0 & a_0
\end{bmatrix}.
$$

The spectral response matrix for $\mathbf{T}_{sf_2}$ is

$$\mathbf{T}_{sf_2} = \mathbf{T}_{f_2}\mathbf{C}_1^* = \begin{bmatrix} 1 & 0 & 0 \\ 0 & 1 & 0 \\ 0 & 0 & 1 \\ 0 & 1 & 0 \\ 0 & 1 & 0 \\ 0 & 0 & 1 \\ 0 & 0 & 1 \\ 0 & 0 & 1 \\ 0 & 0 & 1 \end{bmatrix} \begin{bmatrix} a_0 & a_0 & a_0 \\ a_0 & a_2 & a_1 \\ a_0 & a_1 & a_2 \end{bmatrix} = \begin{bmatrix} a_0 & a_0 & a_0 \\ a_0 & a_2 & a_1 \\ a_0 & a_1 & a_2 \\ a_0 & a_2 & a_1 \\ a_0 & a_2 & a_1 \\ a_0 & a_1 & a_2 \\ a_0 & a_1 & a_2 \\ a_0 & a_1 & a_2 \\ a_0 & a_1 & a_2 \end{bmatrix}.$$

The vector Chrestenson spectra for the two primary outputs can be computed using the relationship $\mathbf{S}_{f_i} = \mathbf{C}_2^*\mathbf{T}_{sf_i}$. $\mathbf{S}_{f_1}$ is computed as follows.

$$\mathbf{T}_{sf_1} = \mathbf{C}_2^*\mathbf{T}_{f_1} = \begin{bmatrix} a_0 & a_0 & a_0 & a_0 & a_0 & a_0 & a_0 & a_0 & a_0 \\ a_0 & a_2 & a_1 & a_0 & a_2 & a_1 & a_0 & a_2 & a_1 \\ a_0 & a_1 & a_2 & a_0 & a_1 & a_2 & a_0 & a_1 & a_2 \\ a_0 & a_0 & a_0 & a_2 & a_2 & a_2 & a_1 & a_1 & a_1 \\ a_0 & a_2 & a_1 & a_2 & a_1 & a_0 & a_1 & a_0 & a_2 \\ a_0 & a_1 & a_2 & a_2 & a_0 & a_1 & a_1 & a_2 & a_0 \\ a_0 & a_0 & a_0 & a_1 & a_1 & a_1 & a_2 & a_2 & a_2 \\ a_0 & a_2 & a_1 & a_1 & a_0 & a_2 & a_2 & a_1 & a_0 \\ a_0 & a_1 & a_2 & a_1 & a_2 & a_0 & a_2 & a_0 & a_1 \end{bmatrix} \begin{bmatrix} a_0 & a_0 & a_0 \\ a_0 & a_1 & a_2 \\ a_0 & a_0 & a_0 \\ a_0 & a_1 & a_2 \\ a_0 & a_1 & a_2 \\ a_0 & a_0 & a_0 \\ a_0 & a_0 & a_0 \\ a_0 & a_0 & a_0 \\ a_0 & a_0 & a_0 \end{bmatrix}$$

$$= \begin{bmatrix} 9 & (6a_0 + 3a_1) & (6a_0 + 3a_2) \\ 0 & (3a_0 + 3a_1) & 3a_1 \\ 0 & 3a_2 & (3a_0 + 3a_2) \\ 0 & (3a_0 + 3a_1) & 3a_2 \\ 0 & 3 & (3a_0 + 3a_1) \\ 0 & 0 & 0 \\ 0 & 3a_2 & (3a_0 + 3a_2) \\ 0 & 0 & 0 \\ 0 & (3a_0 + 3a_2) & 3 \end{bmatrix}$$

Likewise, $\mathbf{S}_{f_2}$ is computed as follows.

$$\mathbf{T}_{sf_2} = \mathbf{C}_2^*\mathbf{T}_{f2} = \begin{bmatrix} a_0 & a_0 & a_0 & a_0 & a_0 & a_0 & a_0 & a_0 & a_0 \\ a_0 & a_2 & a_1 & a_0 & a_2 & a_1 & a_0 & a_2 & a_1 \\ a_0 & a_1 & a_2 & a_0 & a_1 & a_2 & a_0 & a_1 & a_2 \\ a_0 & a_0 & a_0 & a_2 & a_2 & a_2 & a_1 & a_1 & a_1 \\ a_0 & a_2 & a_1 & a_2 & a_1 & a_0 & a_1 & a_0 & a_2 \\ a_0 & a_1 & a_2 & a_2 & a_0 & a_1 & a_1 & a_2 & a_0 \\ a_0 & a_0 & a_0 & a_1 & a_1 & a_1 & a_2 & a_2 & a_2 \\ a_0 & a_2 & a_1 & a_1 & a_0 & a_2 & a_2 & a_1 & a_0 \\ a_0 & a_1 & a_2 & a_1 & a_2 & a_0 & a_2 & a_0 & a_1 \end{bmatrix} \begin{bmatrix} a_0 & a_0 & a_0 \\ a_0 & a_2 & a_1 \\ a_0 & a_1 & a_2 \\ a_0 & a_2 & a_1 \\ a_0 & a_2 & a_1 \\ a_0 & a_1 & a_2 \\ a_0 & a_1 & a_2 \\ a_0 & a_1 & a_2 \\ a_0 & a_1 & a_2 \end{bmatrix}$$

$$= \begin{bmatrix} 9 & (6a_0 + 3a_1) & (6a_0 + 3a_2) \\ 0 & (a_1 + 2a_2) & (5a_0 + a_1) \\ 0 & (5a_0 + a_2) & (2a_1 + a_2) \\ 0 & (a_1 + 2a_2) & (5a_0 + a_1) \\ 0 & (a_0 + 2a_1) & (4a_0 + a_2) \\ 0 & (2a_0 + a_2) & (2a_0 + a_1) \\ 0 & (5a_0 + a_2) & (2a_1 + a_2) \\ 0 & (2a_0 + a_2) & (2a_0 + a_1) \\ 0 & (4a_0 + 2a_1) & (a_0 + 2a_2) \end{bmatrix}$$

□

The spectra of the primary outputs $\mathbf{S}_{f_1}$ and $\mathbf{S}_{f_2}$ are related to the spectrum of the overall MVSN $\mathbf{S}$ in the sense that each primary output spectral coefficient is an outer product factor of the corresponding MVSN spectral coefficient.

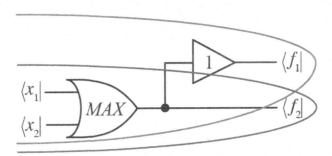

Figure 7.2: Example MVSN a) distributed factored spectral response matrix b) spectral coefficient computation c) output response computation.

To demonstrate this process, we shall use the example MVSN of Chapter 5 depicted in Figure 5.3 in graphical form on the left with the corresponding structural netlist description on the right, and a graphical depiction of the distributed factored form of the transfer matrix at the bottom. The distributed factored form of the spectral response matrix is similar to the bottom portion of Figure 5.3 with the exception that $\mathbf{C}_1^*$ matrices are present at each primary output.

Figure 7.3a depicts the example MVSN as a structural network with switching gate symbols and as a netlist using SystemVerilog syntax. Figure 7.3b is a graphical representation of the corresponding MVSN transfer matrix in distributed factored form and Figure 7.3c is a graphical depiction of the spectral response matrix in distributed factored form. The transfer matrix for the fanout structure is replaced with $\mathbf{I}$ to more easily facilitate traversals. While replacement of $\mathbf{FO}$ with $\mathbf{I}$ is not strictly necessary, it does allow the traversal algorithm to proceed without an intermediate outer product factorization being required. Figure 7.3d likewise contains the spectral response matrix in distributed form with all interconnecting edges annotated by vector values that correspond to the computation of a Chrestenson spectral coefficient.

The particular spectral coefficient computed in Figure 7.3d is the fifth coefficient from the Chrestenson spectrum in natural order since the primary inputs are stimulated with $\langle (12)_3| = \langle 5_{10}|$.

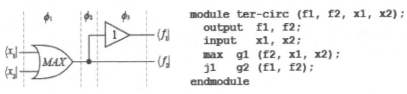

a) Example MVSN and SystemVerilog Structural Netlist

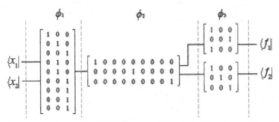

b) Graphical Representation of MVSN Transfer Matrix in Distributed Factored Form

c) Graphical Representation of MVSN Spectral Response Matrix in Distributed Factored Form

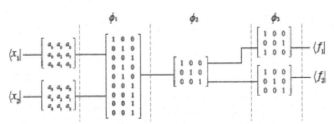

d) MVSN Spectral Response Matrix Annotated with Vectors for Spectral Coefficient Computation

Figure 7.3: Example MVSN a) distributed factored spectral response matrix b) spectral coefficient computation c) output response computation.

CHAPTER 8

Implementation Considerations

For the vector space model to have practical usefulness, it must not incur worse computational complexity in either runtime or storage requirements than would be the case for a conventional switching algebra model. Furthermore, the vector space model must offer some advantages as compared to a corresponding switching algebra implementation. The property of truth table isomorphism guarantees that storage complexity will not be worse than that of conventional switching algebra models. Any data structure capable of representing a switching algebra formulation of a network can also represent the vector space model.

8.1 TRANSFER AND JUSTIFICATION MATRIX REPRESENTATIONS

Two common methods for compactly representing switching functions are cubelists and BDDs. Because the justification matrix is the transpose of the transfer matrix, it is only necessary for one of these structures to be represented for a network of interest.

8.1.1 CUBELIST REPRESENTATIONS OF TRANSFER MATRICES

The .pla format is a common cubelist representation and is described in detail in [19]. Figure 8.1 contains a sample .pla listing for the example network shown in Figure 3.5. For convenience, the example network circuit diagram is also shown in Figure 8.1.

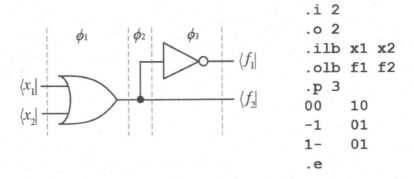

```
.i 2
.o 2
.ilb x1 x2
.olb f1 f2
.p 3
00      10
-1      01
1-      01
.e
```

Figure 8.1: Example network and .pla listing.

The .pla file can be interpreted as a cube list or two-level characterization of a switching circuit modeled with traditional switching algebras. After the initial labeling information on the lines of the file beginning with the character ., two arrays of characters are given containing the symbols 0, 1, and –. The leftmost array represents cubes or product terms and the rightmost array contains information regarding the primary outputs of the network. Here, we use the terminology "input array" to refer to the leftmost array and "output array" to refer to the rightmost array.

Although the symbols 0, 1, and – may appear in either of the input or output arrays, their meaning differs depending on which array they appear in. When a 1 appears in the input array, it represents an instance of the corresponding network input in uncomplemented form while the 0 element represents that variable in complemented form. The appearance of – in the input array denotes that the variable may be considered to be either complemented or uncomplemented and thus corresponds to the $\langle t|$ value in the vector space model. Alternatively, when 0 appears in the output array, it indicates that the switching function produces a 0 value when the cube in the input array on the same line produces a 1, and when 1 appears in the output array, it indicates that the corresponding network output produces a 1 also. The appearance of – in an output array corresponds to the switching algebraic concept of a "don't care" meaning that in a realization of the circuit, the designer is free to assign either 0 or 1 to that output. The use of the – symbol in the input and output arrays often leads to confusion since it has two different meanings depending on which array it appears in. Some degree of compactness is achieved in cube list descriptions as compared to truth tables since the appearance of – in the input array allows for two or more rows in a truth table to be represented as a single row or covering cube in a corresponding .pla description.

Because cubes or product terms may be realized with AND gates and the primary output connectivity represented with OR gates, this form is also a direct representation of a *programmable logic array* or *PLA* two-level circuit form. A direct realization of the .pla listing in Figure 8.1 in two-level form with NOT, AND, and OR gates is provided in Figure 8.2. PLA realizations of switching functions are commonly found in both custom-designed integrated circuits and as programmable structures within FPGAs.

Symbolically, cube lists are often denoted using set notation, and the .pla format is simply one possible representation of the more general notion of a cube list. In the case of the example in Figures 8.1 and 8.2, a covering set for each output f_1 and f_2 is written respectively as $f_1 = \{\bar{x}_1 \bar{x}_2\}$ and $f_2 = x_1, x_2$. Each element of the covering set describes a conjunction (or AND) of switching values and all elements within a set are combined with a disjunctive (or OR) operation to produce the switching function.

Due to truth table isomorphism, the .pla array is alternatively a representation of the transfer matrix of the represented network. The use of the – symbol in the input array causes any particular cube containing a – to represent more than a single row vector in a transfer function representation. In the vector space interpretation of a .pla file, each line can be considered as the specification of a particular vector space mapping. In the example given in Figures 8.1 and 8.2,

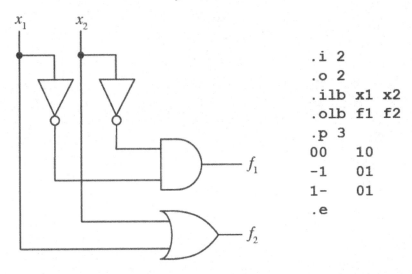

```
.i 2
.o 2
.ilb x1 x2
.olb f1 f2
.p 3
00      10
-1      01
1-      01
.e
```

Figure 8.2: Example `.pla` listing and 2-level implementation.

the three lines containing array information represent the following mappings: 00 10 represents $\langle 00| \rightarrow \langle 10|$, -1 01 represents $\langle t1| \rightarrow \langle 01|$, and 1- 01 represents $\langle 1t| \rightarrow \langle 01|$. To more clearly describe these mappings, Figure 8.3 contains the `.pla` file and the corresponding transfer matrix **T** with the mappings for each row depicted.

8.1.2 BDD REPRESENTATIONS OF TRANSFER MATRICES

The *Binary Decision Diagram* (BDD) representation of switching functions is a widely used technique and can provide significant reductions in storage requirements on average [20]. Although the worst-case complexity remains $O(2^n)$ overall all possible switching functions, ordering and reordering rules provide for the compact representation of many functions. The BDD representing a switching function is isomorphic to a BDD representation of a transfer matrix.

Multi-output networks may be represented with one of various extensions to the BDD model. These include the *Shared Binary Decision Diagram* (SBDD) or the Multi-Terminal Binary Decision Diagram also referred to as the *Algebraic Decision Diagram* (ADD) [23] [52] [21]. Figure 8.4 contains depictions of the SBDD and MTBDD for the example switching network depicted in Figure 3.5.

The transfer matrix interpretation of decision diagrams replaces the switching constants that annotate the terminal vertices (denoted with square boxes) with their corresponding basis vector values ($0 \rightarrow \langle 0|$ and $1 \rightarrow \langle 1|$). The nonterminal vertices are labeled with switching function variables that correspond to switching network primary inputs in the switching algebraic interpretation. The non-terminal vertex annotations and the directed edge labels are replaced with transfer

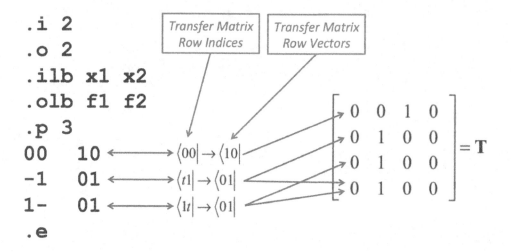

Figure 8.3: Example `.pla` listing and corresponding transfer matrix.

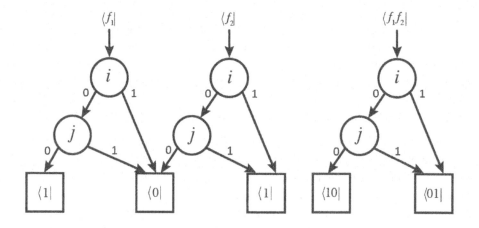

Figure 8.4: Scalar SBDD and MTBDD models for the example switching network.

matrix row and column vector indices when BDDs are interpreted as representations of transfer matrices. Figure 8.5 depicts the corresponding transfer matrix interpretation of the SBDD and MTBDD representation of the example switching network depicted in Figure 3.5.

Due to the identical topological construction of SBDDs and MTBDDs all of the savings in storage requirements for a switching function are realized with the corresponding decision diagram representation. It is often the case that BDD structures are reordered to minimize storage

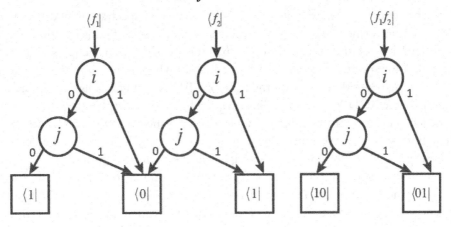

Figure 8.5: Vector SBDD and MTBDD models for the example network transfer function.

requirements. Such reordering is possible for the transfer matrix reinterpretation as well. When the transfer matrix is constructed from a structural netlist representation initially, the arbitrary choice of using the topmost primary input as the first variable in a given order is used in this work. Thus, a reordering of the associated BDD structure corresponds to a permutation matrix applied to the originally parsed transfer matrix. Such permutation matrices may be represented in a variety of structures and accompany BDD representations. The initial ordering corresponds to the permutation matrix $\mathbf{I}$, and subsequent variable interchanges in the BDD are represented through an interchange of rows on the accompanying permutation matrix.

In addition to the use of BDD software for the representation of transfer matrices and vectors, some means of performing linear algebraic operations such as direct and outer products implemented as BDD-based algorithms must be available. Previous work describes how BDD algorithms may be efficiently implemented that perform such linear algebraic operations using BDDs to represent vectors and matrices [12].

8.1.3   STRUCTURAL REPRESENTATIONS OF TRANSFER MATRICES

The other commonly used compact representation of a switching function is the realization of that function as a netlist containing symbols for electronic gates that are modeled as switching function operators. As an example, an integer multiplier is know to produce an exponentially sized BDD regardless of the variable ordering and hence circuits in this class are considered hard cases in the EDA community since their accompanying switching function models are unwieldly. A key advantage of the vector space model presented here is the ability to represent structural net lists as an interconnection of smaller transfer matrices. This allows EDA techniques to be implemented directly over the netlist structure without an intermediate extraction of a switching function representation. An example of this is in the area of spectral methods where switching functions are transformed into alternative bases such as the sets of basis functions known as the Walsh or Reed–Muller functions. A description of the use of decision diagrams as an intermediate representation for spectral methods is described in [7] where switching functions are extracted from a switching network model into the form of a BDD, and graph algorithms are formulated that transform the BDD representation into a *Spectral Decision Diagram* (SDD). Using the vector space model, such spectral transforms may be accomplished directly using a structural netlist representation with no intervening extractions to an intermediate structure. A description of this technique is given in [8] for the case of ternary-valued switching networks.

In using the structural netlist as a representation of a switching network, each gate appearing in the netlist is replaced with a corresponding transfer matrix. Thus the topology of the netlist represents a factorization of the overall network transfer function. The factors are combined with two different multiplicative operators, the direct matrix and the outer product operations. Network elements in parallel are combined with the outer product while serial cascades of the elements are combined with the direct product. In addition to the network element transfer functions, it is necessary to include transfer matrices for cases of fanout, fanin, and crossovers. Figure 8.6 contains a depiction of an example logic network and the corresponding graph depicting the factored transfer matrix.

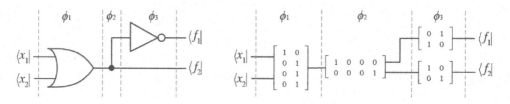

Figure 8.6: Network and graph depicting the factored transfer matrix.

Using the factored transfer matrix graph allows for advantages in implementation of various EDA techniques since algorithmic operations can be applied through a traversal of the factored form structure directly without requiring intermediate extractions of the overall network transfer function. As an example, network simulation can be accomplished within a discrete event simulation framework such as that currently employed in commercial simulation tools by simply changing the values being propagated to be vectors and the simulation of a network element accomplished with a direct vector-matrix product operation. Furthermore, although the network element matrices are small, they can alternatively be replaced with pointers to corresponding cube lists or BDDs.

Justification algorithms within the switching algebraic framework generally employ techniques such as recursive learning [9] or the use of SAT solvers that require first representing the netlist in the form of a list of disjunctive clauses [10] [11]. The factored transfer matrix form shown in Figure 8.6 allows justification to be performed by propagating given output responses from the rightmost side of the factored graph to the leftmost side and multiplying the intermediate vector values with the internal justification matrices in the form of transposed versions of the element transfer matrices.

A generalized version of SAT can be implemented over the factored transfer matrix form. Given a netlist and a set of net values that may appear either internally or upon the primary inputs and outputs, determine a satisfying set of values on all other nets if possible. An iterative invocation of distributed transfer matrix simulation and justification can be performed. Whenever a net is computed to have a value equivalent to the null vector $\langle \varnothing |$ a value in conflict with a previously computed value, the initial input set of assignments can be deemed infeasible, otherwise a solution will result. Furthermore, net values may be computed that contain the value $\langle t |$, thus SAT-all is produced.

8.2   COMPUTING THE TRANSFER MATRIX FROM A STRUCTURAL NETLIST

A previous section described how a transfer matrix may be extracted from a structural description of a switching network such as an HDL netlist through a process of partitioning followed by computation of serial segment transfer matrices, and finally, through a direct multiplication of the segment transfer matrices. Here, we describe details of automating this method for the calculation of a transfer matrix.

8.2.1   SWITCHING NETWORK PARTITIONING

The initial stage requires determination of serial cascade partitions. One method for performing the partitioning phase is to invoke a commonly used method in switching network simulators known as "levelization." Levelization assigns integral-valued level numbers to each line in a structural representation. The primary inputs are initially assigned level numbers of zero; these values are then propagated toward the primary outputs. As a network element is encountered, the level number assigned to the network output is computed by incrementing the maximal-valued level number among all the network element inputs. After all lines contained within the network are assigned level numbers, partition boundaries or vertical "cuts" are determined by choosing all lines with the same level number. To illustrate the levelization-based partitioning technique, we use the small benchmark circuit C17. Figure 8.7 contains the structural representation of C17 with levelization values depicted on each line. Partitioning cuts are depicted along the common level numbers and are shown by dashed lines denoted by χ_i. The circuit partitions are denoted by ϕ_i.

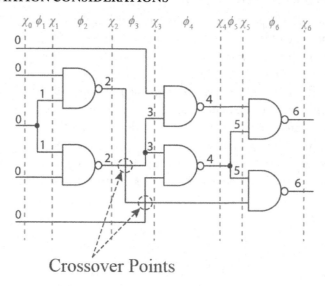

Crossover Points

Figure 8.7: C17 with levelization and partition cuts shown.

8.2.2 COMPUTING THE TRANSFER MATRIX AND CROSSOVER DETECTION

The technique of determining partition transfer matrices has been described in a preceding section and is employed to compute matrices of the network in Figure 8.7. However, an additional set of matrix factors is required to account for situations of crossovers occurring in the structural netlist. One way to account for crossovers is to intersperse permutation matrices, $\mathbf{T}_{\chi_i}$, between the cascade stage matrices that include crossover information. For the situation where no crossovers occur, the permutation matrix is simply $\mathbf{I}$ and may thus be omitted. However, in the case where network lines cross one another within a particular partition, a permutation matrix must be included as a factor in the direct product calculation for the overall transfer matrix. Two crossovers occur in the network depicted in Figure 8.7 and are depicted by dashed circles in the diagram.

For the C17 benchmark, six partitions and seven cuts are identified. Table 8.1 contains the network element transfer matrices organized by partition number.

The overall transfer matrix for C17 is given by Equation 8.1.

$$\mathbf{T}_{C17} = \mathbf{T}_{\chi_0}\mathbf{T}_{\phi_1}\mathbf{T}_{\chi_1}\mathbf{T}_{\phi_2}\mathbf{T}_{\chi_2}\mathbf{T}_{\phi_3}\mathbf{T}_{\chi_3}\mathbf{T}_{\phi_4}\mathbf{T}_{\chi_4}\mathbf{T}_{\phi_5}\mathbf{T}_{\chi_5}\mathbf{T}_{\phi_6}\mathbf{T}_{\chi_6} \quad (8.1)$$

The partition cut transfer matrices, $\mathbf{T}_{\chi_i}$, are determined by the use of values we refer to as "cut indices." For each partition, we assigned values on the immediate right side of each cut line χ_i that are in ascending numerical from top to bottom and where each cut index annotates a network line that crosses a cut or partition line. Within each partition, the cut indices are then

Table 8.1: C17 network element transfer matrices by partition

$\mathbf{T}_{\phi_1}$	$\mathbf{T}_{\phi_2}$	$\mathbf{T}_{\phi_3}$	$\mathbf{T}_{\phi_4}$	$\mathbf{T}_{\phi_5}$	$\mathbf{T}_{\phi_6}$
I	I	I	NA	I	NA
I	NA	FO	NA	FO	NA
FO	NA	I	I	I	
I	I	I			
I					

propagated forward until they reach the input of a network element within the partition. If no network element is present on the cut index, it is propagated to the left side of cut line χ_{i+1}. After propagation of the cut indices within a partition, the cut indices are read from top to bottom and their order indicates if and where crossovers occur. In fact, their order defines a permutation matrix. Also, whenever a cut index is propagated entirely to the next cut line χ_{i+1}, it indicates that the transfer matrix for that network line is the identity $\mathbf{I}$ since a pass-through line is detected.

In the C17 benchmark, all network crossover permutation transfer matrices $\mathbf{T}_{\chi_i} = \mathbf{I}$ except for $\mathbf{T}_{\chi_2}$ since partition ϕ_3 contains two network crossovers. Figure 8.8 depicts C17 with cut indices for partition ϕ_3 shown both as their initial assignment on the right side of cut line χ_2 and their position when propagated. Initially, the cut index order is $(0, 1, 2, 3)$ and after propagation, the order from top to bottom is $(0, 2, 3, 1)$ indicating that two crossovers occurred. The network line with cut index 1 crosses line 2, and then crosses line 3.

After the cut indices have been processed for each partition, the overall expression for the transfer matrix for C17 can be modified. Equation 8.1 is simplified since all but one $\mathbf{T}_{\chi_i} = \mathbf{I}$ and is given in Equation 8.2.

$$\mathbf{T}_{C17} = \mathbf{I}\mathbf{T}_{\phi_1}\mathbf{I}\mathbf{T}_{\phi_2}\mathbf{T}_{\chi_2}\mathbf{T}_{\phi_3}\mathbf{I}\mathbf{T}_{\phi_4}\mathbf{I}\mathbf{T}_{\phi_5}\mathbf{I}\mathbf{T}_{\phi_6}\mathbf{I} = \mathbf{T}_{\phi_1}\mathbf{T}_{\phi_2}\mathbf{T}_{\chi_2}\mathbf{T}_{\phi_3}\mathbf{T}_{\phi_4}\mathbf{T}_{\phi_5}\mathbf{T}_{\phi_6} \qquad (8.2)$$

From the cut index analysis, it is necessary to compute the permutation matrix $\mathbf{T}_{\chi_2}$ that characterizes the fact that a network line (with cut index 1) crosses two lines (with cut indices 2 and 3). Fortunately, multi-line crossovers such as this case can always be modeled as a series of single line crossovers, thus the basic crossover transfer matrix $\mathbf{C}$ as depicted in Figure 3.4 may be used multiple times in the overall calculation for $\mathbf{T}_{\chi_2}$. To illustrate how $\mathbf{C}$ is used multiple times, Figure 8.9 contains a redrawn portion of the C17 circuit containing the portion of partition ϕ_3 where the crossovers occur.

The transfer matrix $\mathbf{T}_{\chi_2}$ can be calculated as

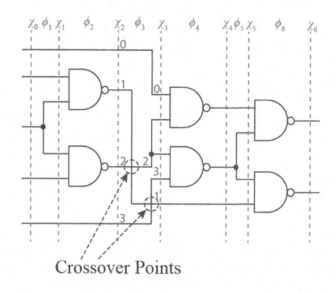

Crossover Points

Figure 8.8: C17 with partition cuts and cut indices shown.

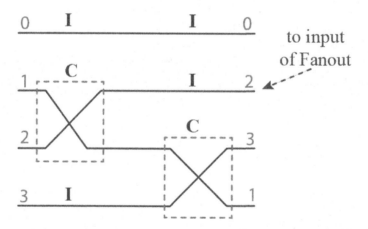

Figure 8.9: Portion of partition ϕ_3 in the C17 circuit with crossovers.

$$\mathbf{T}_{\chi 2} = (\mathbf{I} \otimes \mathbf{C} \otimes \mathbf{I})(\mathbf{I} \otimes \mathbf{I} \otimes \mathbf{C}) = \begin{bmatrix} 1 & 0 & 0 & 0 & 0 & 0 & 0 & 0 & 0 & 0 & 0 & 0 & 0 & 0 & 0 & 0 \\ 0 & 0 & 1 & 0 & 0 & 0 & 0 & 0 & 0 & 0 & 0 & 0 & 0 & 0 & 0 & 0 \\ 0 & 0 & 0 & 0 & 1 & 0 & 0 & 0 & 0 & 0 & 0 & 0 & 0 & 0 & 0 & 0 \\ 0 & 0 & 0 & 0 & 0 & 0 & 1 & 0 & 0 & 0 & 0 & 0 & 0 & 0 & 0 & 0 \\ 0 & 1 & 0 & 0 & 0 & 0 & 0 & 0 & 0 & 0 & 0 & 0 & 0 & 0 & 0 & 0 \\ 0 & 0 & 0 & 1 & 0 & 0 & 0 & 0 & 0 & 0 & 0 & 0 & 0 & 0 & 0 & 0 \\ 0 & 0 & 0 & 0 & 0 & 1 & 0 & 0 & 0 & 0 & 0 & 0 & 0 & 0 & 0 & 0 \\ 0 & 0 & 0 & 0 & 0 & 0 & 0 & 1 & 0 & 0 & 0 & 0 & 0 & 0 & 0 & 0 \\ 0 & 0 & 0 & 0 & 0 & 0 & 0 & 0 & 1 & 0 & 0 & 0 & 0 & 0 & 0 & 0 \\ 0 & 0 & 0 & 0 & 0 & 0 & 0 & 0 & 0 & 0 & 1 & 0 & 0 & 0 & 0 & 0 \\ 0 & 0 & 0 & 0 & 0 & 0 & 0 & 0 & 0 & 0 & 0 & 0 & 1 & 0 & 0 & 0 \\ 0 & 0 & 0 & 0 & 0 & 0 & 0 & 0 & 0 & 0 & 0 & 0 & 0 & 0 & 1 & 0 \\ 0 & 0 & 0 & 0 & 0 & 0 & 0 & 0 & 0 & 1 & 0 & 0 & 0 & 0 & 0 & 0 \\ 0 & 0 & 0 & 0 & 0 & 0 & 0 & 0 & 0 & 0 & 0 & 1 & 0 & 0 & 0 & 0 \\ 0 & 0 & 0 & 0 & 0 & 0 & 0 & 0 & 0 & 0 & 0 & 0 & 0 & 1 & 0 & 0 \\ 0 & 0 & 0 & 0 & 0 & 0 & 0 & 0 & 0 & 0 & 0 & 0 & 0 & 0 & 0 & 1 \end{bmatrix}.$$

The entire set of partition factors for C17 becomes

$$\mathbf{T}_{\phi 1} = (\mathbf{I} \otimes \mathbf{I} \otimes \mathbf{FO} \otimes \mathbf{I} \otimes \mathbf{I})$$
$$\mathbf{T}_{\phi 2} = (\mathbf{I} \otimes \mathbf{NA} \otimes \mathbf{NA} \otimes \mathbf{I})$$
$$\mathbf{T}_{\chi 2} = (\mathbf{I} \otimes \mathbf{C} \otimes \mathbf{I})(\mathbf{I} \otimes \mathbf{I} \otimes \mathbf{C})$$
$$\mathbf{T}_{\phi 3} = (\mathbf{I} \otimes \mathbf{FO} \otimes \mathbf{I} \otimes \mathbf{I})$$
$$\mathbf{T}_{\phi 4} = (\mathbf{NA} \otimes \mathbf{NA} \otimes \mathbf{I})$$
$$\mathbf{T}_{\phi 5} = (\mathbf{I} \otimes \mathbf{FO} \otimes \mathbf{I})$$
$$\mathbf{T}_{\phi 6} = (\mathbf{NA} \otimes \mathbf{NA})$$

The overall form of $\mathbf{T}_{C17}$ is then mathematically modeled as a factored set of network element transfer matrices containing both direct and outer product factors and is expressed in Equation 8.3.

$$\mathbf{T}_{C17} = (\mathbf{I} \otimes \mathbf{I} \otimes \mathbf{FO} \otimes \mathbf{I} \otimes \mathbf{I})(\mathbf{I} \otimes \mathbf{NA} \otimes \mathbf{NA} \otimes \mathbf{I})(\mathbf{I} \otimes \mathbf{C} \otimes \mathbf{I})(\mathbf{I} \otimes \mathbf{I} \otimes \mathbf{C}) \\ (\mathbf{I} \otimes \mathbf{FO} \otimes \mathbf{I} \otimes \mathbf{I})(\mathbf{NA} \otimes \mathbf{NA} \otimes \mathbf{I})(\mathbf{I} \otimes \mathbf{FO} \otimes \mathbf{I})(\mathbf{NA} \otimes \mathbf{NA}) \tag{8.3}$$

The method of parsing, partitioning, and iteratively computing the overall transfer matrix represented as a BDD would typically be accomplished rather than calculating the explicit factored form shown in Equation 8.3 resulting in a BDD representation of the 32×4 transfer matrix $\mathbf{T}_{C17}$ given in Equation 8.4.

$$\mathbf{T}_{C17} = \begin{bmatrix} 1 & 0 & 0 & 0 \\ 0 & 1 & 0 & 0 \\ 1 & 0 & 0 & 0 \\ 0 & 1 & 0 & 0 \\ 1 & 0 & 0 & 0 \\ 0 & 1 & 0 & 0 \\ 1 & 0 & 0 & 0 \\ 1 & 0 & 0 & 0 \\ 0 & 0 & 0 & 1 \\ 0 & 0 & 0 & 1 \\ 0 & 0 & 0 & 1 \\ 0 & 0 & 0 & 1 \\ 0 & 0 & 0 & 1 \\ 0 & 0 & 0 & 1 \\ 1 & 0 & 0 & 0 \\ 1 & 0 & 0 & 0 \\ 1 & 0 & 0 & 0 \\ 0 & 1 & 0 & 0 \\ 1 & 0 & 0 & 0 \\ 0 & 1 & 0 & 0 \\ 0 & 0 & 1 & 0 \\ 0 & 0 & 0 & 1 \\ 0 & 0 & 1 & 0 \\ 0 & 0 & 1 & 0 \\ 0 & 0 & 0 & 1 \\ 0 & 0 & 0 & 1 \\ 0 & 0 & 0 & 1 \\ 0 & 0 & 0 & 1 \\ 0 & 0 & 0 & 1 \\ 0 & 0 & 0 & 1 \\ 0 & 0 & 1 & 0 \\ 0 & 0 & 1 & 0 \end{bmatrix} \tag{8.4}$$

8.3 COMPUTATIONAL RESULTS

Two sets of computations are carried out using the combinational benchmark circuits from [25] and [24]. In both cases, the benchmark circuits were initially converted from their native format (.pla and .isc) into a structural Verilog netlist similar to that depicted in Figure 3.5. In both sets of experiments, the Verilog netlist is parsed into an internal data structure representing the

Table 8.2: Size and computation time of explicit transfer matrices

NAME	IN/OUT	STAGES	PARTITION TIME (ms)	MATRIX TIME (ms)
i3	2/3	3	3.00	5.505
test1	3/3	6	7.28	4.794
xor5	5/1	4	1.73	6.882
majority	5/1	6	11.8	17.71
C17	5/2	7	5.00	22.75
rd53	5/3	6	5.32	10.10
squar5	5/8	9	19.5	922.1
con1	7/2	6	7.09	546.1
rd73	7/3	8	5.01	37.33
radd	8/5	11	12.3	1107
x2	10/7	9	11.4	846.2
cm85a	11/3	11	9.78	1586.2
alu1	12/8	5	8.88	521.9

topological netlist. The internal structure is in the form of a collection of structures where each structure uniquely represents the circuit primary inputs, switching circuit elements, and primary outputs. Each structure also contains pointers that represent topological interconnections of the netlist. This first phase of the experimental software simply parses the structural Verilog description into an equivalent internal structure.

The first computation partitions the network into a serial cascade of network segments with each segment consisting of a parallel set of elements. After the partitioning process, transfer matrices are constructed for each segment and multiplied together to obtain the overall transfer matrix model for the network. After partitioning, the transfer matrices for the two segments closest to the primary inputs are formed first and then they are multiplied together. Next, the segment transfer matrix for the third section is formed and then it is multiplied with the previously obtained matrix product. In this manner, the overall transfer matrix is iteratively formed by forming a segment transfer matrix and then multiplying it with the accumulated product matrix. After all segments have been processed, the product matrix is the overall network transfer matrix. In the first set of experiments, matrices are represented as explicit two-dimensional arrays and thus become exponentially large with respect to the number of primary inputs and primary outputs. Table 8.2 contains the results of this experiment. The first column contains the benchmark circuit name, the second column contains the number of primary inputs and outputs, the third column is the number of stages found during the partitioning process, the fourth column contains the time required to partition the network, and the fifth column contains the total time required to formulate the transfer matrix.

Table 8.3: Size and computation time of transfer matrices using the BDD data structure

NAME	IN/OUT	MATRIX SIZE (KB)	MATRIX TIME (ms)
C880	60/26	18513.56	60
C1355	41/32	2876.28	50
C1908	33/25	1193.72	90
C3540	50/22	40409.64	3940
apex7	49/37	26.39	133.1
dalu	75/16	51064.39	563.7
x4	94/71	79.84	8.264
apex5	117/88	42.41	296.5
ex4	128/28	143.72	113.9
frg2	143/139	102.63	306.9
i2	201/1	8.36	4.647

The second set of computations parses the Verilog file and builds the transfer matrix utilizing the BDD data structure. Using the BDD structure greatly reduces both the required storage and runtime required to construct the transfer matrix.

CHAPTER 9

Summary

The vector space model allows for the formulation of Boolean problems as linear algebraic problems and thus the rich set of results in this area of mathematics may be used. Popular computational tools such as MATLAB$^{©}$ may be used to solve Boolean problems such as simulation and justification of switching networks. One advantage of this approach is that modern EDA software licensing is very expensive and can cost as much as several $100 thousands of dollars per year. This high cost can be a challenge for startup companies and academic or research settings. The use of the vector space model can allow EDA tasks to be reformulated requiring only the use of low-cost or open source linear algebraic software.

The linear algebraic framework provides alternative insight into the nature, analysis, and design of switching networks. While it is not expected that switching theory models will be abandoned, it is possible that the linear algebraic framework may lead to a new class of techniques for switching network and information processing. The previous chapters have shown that simulation, justification, and spectral computations can be applied with as much efficiency as that in current use based on switching theory algebras.

The general SAT-all problem can be formulated as a series of simulation and justification operations allowing all primary input, primary output, and internal net values to be found given a sufficient subset of the values initially. An iterative application of simulation and justification computations can be used to determine assignment values for all nets within a given netlist based upon an initial subset of values. The calculations provide all satisfying net values for each computation, and infeasible initial assignments can be quickly determined when conflicting or null assignments are computed.

It is shown that the computational complexity of the linear algebraic formulations do not incur worse penalties than those of Boolean formulations and, in some cases, advantages may be present in using this alternative approach. Although the theoretical computational complexities for Boolean problems remain, heuristic approaches for their solution may benefit when linear algebraic methods are employed as compared to traditional Boolean switching algebraic solutions. The computational results indicate that the vector space models do provide a practical alternative to switching algebraic solutions and thus are competitive candidates for use in solving Boolean problems. It is not expected that the community will abandon the traditional approaches grounded in switching theory; however, it is encouraging that some classes of Boolean problems may be more conveniently modeled as problems in the vector space.

Future investigations include reframing other common Boolean problems in the algebraic framework and evaluating their effectiveness. Additionally, we intend to utilize these methods for other types of information processing systems such as fuzzy logic and general MVL systems of varying radix value. One advantage of the vector space model is that it provides a unifying mathematical framework among the conventional binary, MVL, fuzzy, and quantum logic communities. The differences among these fields are the dimension of the vectors representing the atomic information datum and the type of components used. Boolean and discrete-valued MVL systems utilize vectors of dimension r^n where r is the radix value and n is a member of the set $\mathbb{B} = \{0, 1\}$, while fuzzy logic vector components are real-valued and members of $\mathbb{R}$, and quantum logic vector components are complex-valued and members of the set $\mathbb{C}$. This unifying structure may allow for more convenience in evaluating Boolean problems in alternative information processing paradigms.

Bibliography

[1] G. Boole, *An Investigation of the Laws of Thought on Which are Founded the Mathematical Theories of Logic and Probabilities*, reissued by Cambridge Press (2009), Ireland, 1854. DOI: 10.1017/CBO9780511693090. 1

[2] C. E. Shannon, *A Symbolic Analysis of Relay and Switching Circuits*, M. I. T. M. S. Thesis, Boston, MS., 1937. 1

[3] C. E. Shannon, "A Mathematical Theory of Communication," *Bell System Technical Journal*, vol. 27, 1948, pp. 379–423. DOI: 10.1002/j.1538-7305.1948.tb01338.x. 1

[4] J. von Neumann, *Mathematical Foundations of Quantum Mechanics*, reissued by Princeton University Press (1996), Princeton, N.J., 1932. 2

[5] P. A. M. Dirac, "A New Notation for Quantum Mechanics," *Proc. of the Cambridge Philosophical Society*, vol. 35, 1939, p. 416. DOI: 10.1017/S0305004100021162. 2, 8

[6] E. H. Moore, "On the Reciprocal of the General Algebraic Matrix," *Bulletin of the American Mathematical Society*, vol. 26, no. 9, 1920, pp. 394–395. 2, 30

[7] M. A. Thornton, R. Drechsler, and D. M. Miller, *Spectral Methods in VLSI CAD*, Kluwer Academic Publishers, Boston, Massachussetts, July 2001. DOI: 10.1007/978-1-4615-1425-1. 55, 61, 100, 124

[8] M. A. Thornton and T. W. Manikas, "Spectral Response of Ternary Logic Netlists," in proc. *IEEE International Symposium on Mulitple-Valued Logic* (ISMVL), May 24–25, 2013, pp. 109–116. DOI: 10.1109/ISMVL.2013.52. 124

[9] W. Kunz and P. R. Menon, "Multi-level Logic Optimization by Implication Analysis", in proc. *IEEE/ACM International Conference on Computer-aided Design* (ICCAD), November 6–10, 1994, pp. 6–13. DOI: 10.1109/ICCAD.1994.629735. 124

[10] J. P. Marques-Silva and K. A. Sakallah, "GRASP-A New Search Algorithm for Satisfiability," in proc. *IEEE/ACM International Conference on Computer-aided Design* (ICCAD), November 10–14, 1996, pp. 220–227. DOI: 10.1109/ICCAD.1996.569607. 124

[11] M. W. Moskewicz, C. F. Madigan, Y. Zhao, L. Zhang, and S. Malik, "Chaff: Engineering an Efficient SAT Solver," in proc. *IEEE/ACM Design Automation Conference* (DAC), 2001. DOI: 10.1145/378239.379017. 124

[12] E. M. Clarke, M. Fujita, P. C. McGeer, K. McMillan, J. C.-Y. Yang, and X. Zhao, "Binary Decision Diagrams: An Efficient Data Structure for Matrix Representation," in proc. *IEEE Int. Workshop on Logic Synthesis*, pp. 1–15, 1993. 123

[13] H. Savoj, *Don't Cares in Multi-level Network Optimization*, Ph.D. dissertation, University of California, Berkeley, 1992. 11

[14] C. T. Chen, Linear System Theory, Holt, Rinehart, and Winston, 1984. 14

[15] M. A. Nielsen and I. L. Chuang, *Quantum Information and Quantum Computing*, Cambridge University Press, 2000. 14

[16] IEEE Design Automation Standards Committee, "IEEE Standard Multivalue Logic System for VHDL Model Interoperability," IEEE Press, New Jersey, 1993. 11, 37

[17] IEEE Design Automation Standards Committee, "1364-2001-IEEE Standard Verilog Hardware Description Language," IEEE Press, New Jersey, 2001. 11, 37

[18] R. Landauer, "Irreversibility and Heat Generation in the Computing Process," *IBM Jour. Res. Develop.*, vol. 5 no. 3, 1961, (reprinted in *IBM Jour. Res. Develop.*, vol. 44 no 1/2, January/March, pp. 265–269, 1961, 2000). DOI: 10.1147/rd.53.0183. 29

[19] espresso(5) man page, "Berkeley PLA Tools," *man page released with Berkeley PLA tools*. 119

[20] R. E. Bryant, "Graph-based Algorithms for Boolean Function Manipulation," *IEEE Trans. on Computers*, vol. C-35, no. 8, 1986, pp. 677–691. DOI: 10.1109/TC.1986.1676819. 121

[21] R. I. Bahar, E. A. Frohm, C. M. Ganoa, G. D. Hachtel, E. Macii, A. Pardo, and F. Somenzi, "Algebraic Decision Diagrams and Their Applications," in proc. *IEEE/ACM International Conference on Computer-aided Design* (ICCAD), 1993. DOI: 10.1109/IC-CAD.1993.580054. 121

[22] E. Clarke, M. Fujita, and X. Zhao, "Applications of Multi-Terminal Binary Decision Diagrams," *Technical Report*, Carnegie Mellon University, CMU-CS-95-160, April, 1995. 100

[23] S.-I. Minato, N. Ishiura, and S. Yajima:, "Shared Binary Decision Diagram with Attributed Edges for Efficient Boolean Function Manipulation," in proc. *IEEE/ACM Design Automation Conference* (DAC), 1990. DOI: 10.1109/DAC.1990.114828. 121

[24] F. Brglez and H. Fujiwara, "A Neutral Netlist of 10 Combinational Benchmark Circuits and A Target Translator in FORTRAN," in proc. *IEEE International Symposium on Circuits and Systems* (ISCAS), pp. 663–698, June 1985. 130

[25] S. Yang, "Logic Synthesis and Optimization Benchmarks," version 3.0, *Technical Report*, Microelectronics Center of North Carolina, 1991. 130

[26] D. M. Miller and M. A. Thornton, *Multiple Valued Logic: Concepts and Representations.* Morgan & Claypool Publishers, 2007. DOI: 10.2200/S00065ED1V01Y200709DCS012. 35, 36, 40

[27] M. A. Thornton, "A Transfer Function Model for Ternary Logic Circuits," *Proceedings. IEEE International Symposium on Multiple-Valued Logic*, pp. 103–108, 2013. DOI: 10.1109/ISMVL.2013.10. 35, 41

[28] M. A. Thornton and J. Dworak, "Ternary logic network justification using transfer matrices," *Proceedings. IEEE International Symposium on Multiple-Valued Logic*, pp. 310–315, 2013. DOI: 10.1109/ISMVL.2013.57. 35

[29] M. A. Thornton and T. W. Manikas, "Spectral response of ternary logic netlists," *Proceedings. IEEE International Symposium on Multiple-Valued Logic*, 109–116, 2013. DOI: 10.1109/ISMVL.2013.52. 35

[30] E. L. Post, "Introduction to a General Theory of Elementary Propositions," *Amer. J. Math.*, vol. 43, pp. 163–185, 1921. DOI: 10.2307/2370324. 35, 101

[31] T. Sasao, *Switching Theory for Logic Synthesis.* Kluwer Academic Publishers, 1999. DOI: 10.1007/978-1-4615-5139-3. 2

[32] J. Roth, "Diagnosis of automata failures: A calculus and a method," *IBM Journal of Research and Development*, vol. 10, no. 4, pp. 278–291, 1966. DOI: 10.1147/rd.104.0278. 29

[33] P. Goel, "An implicit enumeration algorithm to generate tests for combinational logic circuits," *IEEE Trans. Comput.*, vol. C-30, no. 3, pp. 215–222, 1981. DOI: 10.1109/TC.1981.1675757. 29

[34] H. Fujiwara and T. Shimono, "On the acceleration of test generation algorithms," *IEEE Trans. Comput.*, vol. C-32, no. 12, pp. 1137–1144, 1983. DOI: 10.1109/TC.1983.1676174. 29

[35] W. Kunz and D. K. Pradhan, "Recursive learning: An attractive alternative to the decision tree for test generation in digital circuits," *IEEE Trans. Comput.-Aided Design Integr. Circuits Syst.*, vol. 13, no. 9, pp. 1143 –1158, Sep. 1994. DOI: 10.1109/TEST.1992.527905. 30

[36] M. A. Thornton, "Spectral Analysis of Digital Logic using Circuit Netlists," *International Conf. on Computer-Aided Systems Theory*, pp. 414–415, 2011. 58

[37] R. Stanković, and J. Astola, *Spectral Interpretation of Decision Diagrams*, Springer-Verlag Publishers, 2003. 55, 61

[38] S. L. Hurst, D. M. Miller, and J. C. Muzio *Spectral Techniques in Digital Logic*, Academic Press, 1985. 55

[39] M. G. Karpovsky, *Finite Orthogonal Series in the Design of Digital Devices*, John Wiley Publishers, 1976. 55, 102, 104

[40] M. G. Karpovsky, Stanković, and J. Astola, *Spectral Logic and its Applications for the Design of Digital Devices*, John Wiley & Sons, 2008. DOI: 10.1002/9780470289228. 55

[41] B. Falkowski and M. A. Perkowski, "One More Way to Calculate the Hadamard-Walsh Spectrum for Completely and Incompletely Specified Boolean Functions," *Int. Jour. of Electronics*, vol. 69, no. 5, pp. 595–602, 1990. DOI: 10.1080/00207219008920344. 55

[42] B. Falkowski and M. A. Perkowski, "One More Way to Calculate the Generalized Reed-Muller Expansions of Boolean Functions," *Int. Jour. of Electronics*, vol. 71, no. 3, pp. 385–396, 1991. DOI: 10.1080/00207219108925484. 55

[43] D. M. Miller, "Graph Algorithms for the Manipulation of Boolean Functions and their Spectra," *Congressus Numerantium*, 57, pp. 177–199, 1987. 55, 61

[44] R. Drechsler and M. A. Thornton, "Computation of Spectral Information form Logic Netlists," *IEEE Int. Symp. on Multiple-Valued Logic*, pp. 53–58, 2000. DOI: 10.1109/ISMVL.2000.848600. 55, 56

[45] R. Krenz, E. Dubrova, and A. Kuehlmann, "Fast Algorithm for Computing Spectral Transforms of Boolean and Multiple-Valued Functions on Circuit Representation," *IEEE Int. Symp. on Multiple-Valued Logic*, pp. 334–339, 2003. DOI: 10.1109/ISMVL.2003.1201426. 55, 56

[46] R. E. Blahut, *Fast Algorithms for Digital Signal Processing*, Addison Wesley, 1985. 60

[47] I. J. Good, "The Interaction Algorithm and Practical Fourier Analysis," *J. Royal Statis. Soc., Ser. B*, 20, pp. 361–375, 1958; addendum, 22, pp. 372–375, 1960. 60

[48] J. W. Cooley and J. W. Tukey, "An Algorithm for the Machine Computation of Complex Fourier Series," *Math. Comp.*, 19, pp. 297–301, 1965. DOI: 10.1090/S0025-5718-1965-0178586-1. 60

[49] M. A. Thornton and R. Drechsler, "Spectral Decision Diagrams using Graph Transformations," *IEEE/ACM Conference on Design, Automation, and Test in Europe*, pp. 713–717, 2001. 61

[50] K. D. Heidtmann, "Arithmetic Spectrum Applied to Fault Detection for Combinational Networks," *IEEE Trans. on Computers*, vol. C-40, no. 3, 1991. DOI: 10.1109/12.76409. 100

[51] S. B. K. Vrudhula, M. Pedram, and Y.-T. Lai, "Edge Valued Binary Decision Diagrams," Chapter in *Representation of Discrete Functions*, editors T. Sasao and M. Fujita, Kluwer Academic Publishers, pp. 109–132, 1996. DOI: 10.1007/978-1-4613-1385-4. 100

[52] E. M. Clarke, M. Fujita, and X. Zhao, "Hybrid Decision Diagrams-Overcoming the Limitations of MTBDDs and BMDs," *IEEE International Conference on Computer Aided Design*, pp. 159–163, 1995. DOI: 10.1109/ICCAD.1995.480007. 121

[53] S. K. Kumar and M. A. Breur, "Probabilistic Aspects of Boolean Switching Functions via a New Transform," *Journal of the ACM*, vol. 28, no. 3, pp. 502–520, 1981. DOI: 10.1145/322261.322268. 100

[54] H. E. Chrestenson, "A Class of Generalized Walsh Functions," *Pacific Journal of Mathematics*, vol. 5, pp. 17–31, 1955. DOI: 10.2140/pjm.1955.5.17. 101

[55] M. Amoui, D. Große, M. A. Thornton and R. Drechsler , "Evaluation of Toggle Coverage for MVL Circuits Specified in the SystemVerilog HDL," *IEEE Int. Symp. on Multiple-Valued Logic*, Session 8B, paper 2 (on CD), 2007. DOI: 10.1109/ISMVL.2007.19. 102

[56] C. Moraga , "Complex Spectral Logic," *IEEE Int. Symp. on Multiple-Valued Logic*, pp. 149–156, 1978. 102, 104

Author's Biography

MITCHELL A. THORNTON

Mitch Thornton is a Professor of Computer Science and Engineering and a Professor of Electrical Engineering at Southern Methodist University. Additionally, he serves as the Technical Director of the Darwin Deason Institute for Cyber Security, also at SMU. He was designated as the J. Lindsey Embrey Chair in Computer Science and Engineering in 2004 and as a Gerald Ford Research Fellow at SMU in 2005. His industrial experience includes employment at Amoco Research Center, E-Systems, Inc. (now L-3 Communications) and the Cyrix Corporation where he held a variety of engineering and technical positions. He has practiced as an independent professional engineer since 1993. His practice areas include all aspects of digital systems design and analysis, computer architecture, computer systems security, and embedded systems. He has published more than 200 technical articles, authored or co-authored five books, and is a named inventor on two U.S. patents and three patents pending. Mitch has consulted with and performed sponsored research for a variety of government and industrial organizations. His research interests include hardware computer security, electronic design automation, disaster and fault tolerance, and emerging technology. He is a licensed professional engineer in the states of Texas, Arkansas, and Mississippi. Mitch received the Ph.D. in computer engineering from SMU, M.S. in computer science from SMU, M.S. in electrical engineering from the University of Texas at Arlington, and B.S. in electrical engineering from Oklahoma State University.

Index

Printed in the United States
by Baker & Taylor Publisher Services